Quasi-Least Squares Regression

MONOGRAPHS ON STATISTICS AND APPLIED PROBABILITY

General Editors

F. Bunea, V. Isham, N. Keiding, T. Louis, R. L. Smith, and H. Tong

Monographs on Statistics and Applied Probability 132

Quasi-Least Squares Regression

Justine Shults

University of Pennsylvania, Philadelphia, USA

Joseph M. Hilbe

Jet Propulsion Laboratory
California Institute of Technology, USA

and

Arizona State University, USA

CRC Press
Taylor & Francis Group
Boca Raton London New York

CRC Press is an imprint of the
Taylor & Francis Group, an **informa** business

A CHAPMAN & HALL BOOK

First published in paperback 2024

First published 2014 by Chapman and Hall/CRC Press

Published 2019 by CRC Press
2385 NW Executive Center Drive, Suite 320, Boca Raton FL 33431

and by CRC Press
4 Park Square, Milton Park, Abingdon, Oxon, OX14 4RN

CRC Press is an imprint of Taylor & Francis Group, LLC

Library of Congress Cataloging-in-Publication Data

Shults, Justine.
   Quasi-least squares regression / Justine Shults, Joseph M. Hilbe.
      pages cm. -- (Chapman & Hall/CRC monographs on statistics & applied
      probability)
   "A CRC title."
   Includes bibliographical references and index.
   ISBN 978-1-4200-9993-5 (hardcover : alk. paper) 1. Least squares. 2. Regression
   analysis--Mathematical models. I. Hilbe, Joseph M., 1944- II. Title.

QA275.S48 2014
519.2'3--dc23   2013047324

ISBN: 978-1-4200-9993-5 (hbk)
ISBN: 978-1-03-292694-0 (pbk)
ISBN: 978-0-429-13227-8 (ebk)

DOI: 10.1201/b16446

Contents

Preface

Overview

The generalized estimating equation (GEE) approach (Liang and Zeger, 1986) has become one of the most popular methods for modeling longitudinal and clustered data, particularly in the area of biostatistics. Its popularity stems from its usefulness. GEE extends generalized linear models (GLM) so that correlated data might be appropriately modeled. This is accomplished by relaxing the assumption of independence of measurements, which is violated when the data are longitudinal or in a clustered or panel form.

Quasi-least squares (SQLS) is a two stage computational approach for estimation of the correlation parameters that is in the framework of GEE. Stage one of QLS was proposed for balanced data by Chaganty (1997) and for unbalanced and unequally spaced data by Shults and Chaganty (1998), while stage two was proposed by Chaganty and Shults (1999). As we shall see in this monograph, QLS extends the application of GEE by allowing for easier consideration of new patterns of association, improved estimation of the regression and correlation parameters in some situations, and an alternative option for analysis should GEE fail to converge.

We first describe GLM and GEE, which comprise the foundation for QLS. We then discuss various limitations of GEE and provide a thorough description of the theoretical underpinnings of the QLS approach that highlights the relative advantages of QLS. We describe how to implement the standard correlation structures that are available for GEE, followed by newer structures that are plausible for unequally or equally spaced longitudinal data, familial data, and data with multiple sources of correlation. The latter structures are not currently available in the major software implementations of GEE.

Chapter 7 considers a recent topic of debate in the statistical literature, which is whether semi-parametric methods such as QLS and GEE are even appropriate for the analysis of discrete data. The problem is that although semi-parametric approaches only require specification of the first two moments of the underlying distribution, for discrete data, it is possible to obtain estimates for which there can be no valid distribution with the estimated means and correlations. We discuss the theoretical and practical implications of this issue in the context of logistic regression for longitudinal data. Our assessment includes a comparison between QLS and the recently developed first-order Markov maximum-likelihood (MARK1ML) approach for analysis of longitudinal binary data.

Chapter 8 offers an evaluation of methods for selection of the best-fitting correlation structure. This is important because the enhanced ability of QLS to implement new patterns of association requires accompanying approaches to choose between the structures. We implement working correlation structures that were not considered in prior comparisons of methods for selection of a working structure. Our comparisons utilize a recently developed approach for simulation of discrete data with decaying product correlations.

The final chapter provides a discussion of sample size calculation, and a worked example demonstrating the application of QLS for estimation of model parameters, assessment of the fit of several candidate working correlation structures, and sample size calculation for planning a future study.

Intended Audience

This text provides the means to conduct an improved GEE analysis for study designs that are commonly encountered in research and should therefore be a valuable resource for researchers who would like to appropriately account for the correlation in their data using an approach that is straightforward to apply. However, we also provide a thorough description of the theoretical underpinnings of QLS so that our intended audience for this book includes both statisticians (and perhaps statistically sophisticated non-statisticians) who are interested in applying QLS in their analyses, in addition to statisticians who might be interested in working on some of the open research problems in this area. Those who are primarily interested in implementing QLS and GEE could skim over the theoretical developments, and head straight to the worked examples that are offered at the end of each chapter.

With the approval of our editor Rob Calver, the first author (Shults) used the manuscript version of this book as the text for an advanced elective course for students in the PhD program in biostatistics at the University of Pennsylvania. Her positive experience using the manuscript for the course indicates that this monograph may be suitable as a text for advanced graduate students who are enrolled in a PhD program in biostatistics or in statistics. Suggested prerequisites for the course include linear models, with some knowledge of longitudinal data analysis (and in particular GEE) being helpful. However, if the instructor focuses on the worked examples in each chapter, this monograph could serve as a supplemental, or even primary, text for a course on analysis of correlated data, or perhaps for a course on improved GEE-based analyses of longitudinal and clustered data. Problems focusing on theory and applications are included at the end of each chapter. The instructor may select particular problems that are most appropriate for his or her students.

Software

We have primarily used Stata statistical software, version 13.0 (2013), for demonstrating example modeling output. User-authored software using Stata's proprietary programming language was developed by Shults et al. (2007) for general QLS estimation; extensions of this software to R, SAS, and MATLAB® are also available (Xie and Shults, 2009; Kim and Shults, 2010; Ratcliffe and Shults, 2008).

Readers are also encouraged to refer to the website for this monograph, where we will post updates and accompanying code and guidelines to replicate many of the examples in this book in Stata, R, SAS, and MATLAB, https://dbe.med.upenn.edu/biostat-research/Book-QLS. Additional information and errata will also be available on Professor Shult's Be-Press Selected Works Site, http://works.bepress.com/justine_shults. Resources will also be provided on the publisher's website for the book, http://www.crcpress.com/product/isbn/9781420099935.

Acknowledgments

We gratefully acknowledge the National Cancer Institute of the National Institutes of Health, which provided the funding for Professor Shults for the Longitudinal Analysis for Diverse Populations Project (LADP Project R01-CA096885). The goal of the LADP project was to develop more efficient and cost-effective methods for analysis of longitudinal studies using quasi-least squares (QLS), with special emphasis on studies in diverse populations that included community-based interventions. Many of the results in this text stem from the LADP project that is described on the website for the project, https://dbe.med.upenn.edu/biostat-research/ladp.

We also wish to acknowledge the assistance and insight from a number of individuals who have been associated with our work. J. Shults is extremely grateful to Ardythe Morrow for first encouraging her and her dissertation advisor, N. Rao Chaganty, to learn more about GEE. She also thanks Professor Chaganty and his former students Genming Shih, Deepak Mav, Yihao Deng, and Roy Sabo, for their research on QLS and GEE. She thanks Professor Joe Gastwirth for being a wonderful mentor over the years. J. Shults also thanks James Hardin for developing the Stata command **xtgee**; she used **xtgee** in the **xtqls** command to solve the GEE estimating equation for the regression parameter at the current QLS estimate of the correlation parameter.

J. Shults is also extremely grateful to her many colleagues at the University of Pennsylvania, and in particular to those who worked with her on the LADP project and on research concerning QLS, including her close colleague and good friend Sarah Ratcliffe, who developed MATLAB® software for QLS for longitudinal data and for data with multiple sources of correlation. She is also grateful to Mary Leonard, a pediatric nephrologist and clinical epidemiologist with whom she has worked closely on statistical issues involved in the longitudinal analysis of bone density, structure, and strength since 1999. She is also grateful to her fellow co-director of the pediatric section in the Department of Biostatistics, Russell Localio, for his generous advice and expertise on methods for analysis of longitudinal data. She is also very grateful to Jimbo Chen, Scarlett Bellamy, Carissa A. Mazurick, and Richard Landis.

As a faculty member in the Department of Biostatistics and Epidemiology at the University of Pennsylvania, J. Shults has had the honor of working with many excellent graduate students. She thanks Jichun Xie for her work on the development of the R package **qlspack** and on the implementation of QLS for familial data that is featured in Chapter 6; Matthew White, for his work on comparison of methods for selection of a correlation structure for GEE and QLS; Seunghee Beck, Chia-

Hao Wang, Xiaoying Wu, Jiwei He, Arwin Thomassen, and Yimei Li for exploring issues related to QLS in their master's theses in biostatistics at the University of Pennsylvania. She also thanks former student Wenguang Sun for his work on the comparison of QLS with other methods that utilize unbiased estimating equations for the regression parameter. She also thanks Qian Wu (Vicky) for useful comments she provided on this text when she was enrolled in the advanced elective course on QLS at Penn.

J. Shults is also very grateful to her first PhD student, Hanjoo Kim, for his work on extending QLS for unbalanced data with multiple sources of correlation and for his development of software for QLS in SAS for data with one and multiple sources of correlation. She is thankful for Hanjoo's continued enthusiasm to collaborate on research, some of which is cited in this text. She also thanks her second PhD student, Matthew W. Guerra, in particular for his input regarding the comparisons with MARK1ML, a recently developed approach that will be featured in their forthcoming book on logistic regression for correlated data (Guerra and Shults, 2014).

J. Shults would also like to acknowledge two faculty members who sadly, were lost recently. She is extremely grateful to former Professor Dayanand Naik for his ground-breaking work on multivariate analysis and for setting a wonderful example in terms of kindness and strength of character. She is also exceedingly grateful to former Professor Thomas Ten Have for his guidance and collaboration on issues related to QLS and for also setting a wonderful example as an outstanding researcher with integrity.

J. Hilbe wishes to acknowledge his long working relationship and friendship with Professor James Hardin, with whom he has co-authored five texts on GLM and GEE, as well as a number of encyclopedia and journal articles. Texts include Hardin & Hilbe, *Generalized Estimating Equations* (2003, 2013), Chapman & Hall/CRC, and *Generalized Linear Models and Extensions* (2001, 2007, 2012), Stata Press, CRC Press. Professors Hardin and Hilbe co-authored the current version of Stata's **glm** command in 2001, which was a revision of Hilbe's initial version **glm**, first published in January 1993.

Both authors would also like to thank several additional people, whose valuable insights are much appreciated. They are extremely grateful to their patient and supportive editor Rob Calver, who has provided excellent advice and guidance throughout the writing process. They really appreciate all the time he put into working with them on this monograph, and enthusiastically recommend him as an editor to other statisticians who might contemplate writing a text.

In addition, special thanks are due to our external reviewers. Professor Preisser's detailed comments led to many improvements throughout the monograph, in addition to a new section on the characteristics of clustered and longitudinal data (Section 5.1). Overall, his reviews were extremely detailed and thoughtful, and we are very grateful for his efforts. Professor Adrian Barnett also provided thoughtful and detailed reviews regarding the content that led to substantial improvements in the discussions and to important changes in the text, for example to improve the clarity of explanations. Thanks are also due to Shashi Kumar of Cenveo Publisher Services, who provided thoughtful advice regarding LaTeX. We also thank our project editor Karen Simon for her advice and editing when the book was in production.

We also wish to acknowledge and thank our spouses and children. The first author's primary responsibility for writing the chapters came at the expense of time spent with her family. She is therefore exceedingly grateful to her wonderful husband and children, Chuck, Chucky, and Erika, in addition to her former dogs, Bud (who almost made it to the age of 16) and Noble (a sweet pit bull originally from the Philadelphia SPCA, who recently lost his battle with lymphoma). The second author also wishes to thank the members of his family who once again had to lose time they would have otherwise spent together. To his wife Cheryl, his children and grandchildren, and to his constant companion Sirr, a small white Maltese, he expresses his deepest gratitude.

MATLAB® is registered trademark of The MathWorks, Inc. For product information, please contact:

The MathWorks, Inc.
3 Apple Hill Drive
Natick, MA 01760-2098 USA
Tel: 508-647-7000
Fax: 508-647-7001
E-mail: info@mathworks.com
Web: www.mathworks.com

Part I

Introduction

Chapter 1

Introduction

Quasi-least squares (QLS) is a method for analysis of correlated data that is in the framework of the generalized estimating equation (GEE) approach introduced by Liang and Zeger (1986). Prior to delving into the technical details, in this introductory chapter we provide an overview of GEE and of QLS. We also describe several datasets with characteristics that are commonly encountered in studies that yield correlated data. We will refer back to these datasets throughout the book, as we describe QLS, demonstrate its implementation, and make comparisons with other popular approaches for the analysis of correlated data.

1.1 GEE and QLS for Analysis of Correlated Data

Correlated data are often encountered in research. For example, in a study that assesses birth-weight of rats within litters, we might anticipate that rats will be more similar if they are litter-mates; as a result, the intra-litter correlations will be non-zero. Or, if we randomly select one rat from each litter and compare the average change in weight over time between groups (e.g., rats who were longer versus shorter at birth), then we might anticipate that weights measured on the same rat will be more similar than weights measured on different rats; as a result, the intra-rat correlations will be non-zero.

Each of the studies just described involves *clusters* of measurements, within which we expect measurements to be correlated, but between which an assumption of independence is reasonable. In the birth-weight study the clusters comprise single measurements collected on rats within litters, while in the weight-change study the clusters comprise repeated measurements collected on each of a group of independent rats. In Chapter 6 we will also consider methods for analysis of multi-level correlated data, which could involve repeated measurements collected on rats within litters; for this example the clusters comprise repeated measurements within each litter.

QLS and GEE are both relatively straightforward approaches for analysis of correlated data and, as such, can be useful for the analysis of data from medical (and other) studies, when the goal is to compare means between groups, and especially when the analysts have at least a working knowledge of linear, logistic, or Poisson regression.

Longitudinal Weight Loss Study: To provide a context for our discussion, in this section we consider a longitudinal study for promotion of weight loss. Goals of the study include comparing the difference between two treatment groups in terms of: (1) the absolute weight change since baseline, and (2) whether a subject has lost at least five percent of their body weight since baseline.

In this study, the comparisons of interest involve average change between the groups, so that a population-averaged approach is appropriate for analysis of the data from this trial. The *population-averaged* approach is in contrast to a *subject-specific* approach that would yield individual level predictions. For example, a population-averaged approach might be used to compare average weight change between intervention versus control subjects, while a subject-specific approach (Diggle et al., 2002; Hardin and Hilbe, 2012; Zeger et al., 1988) could be used to predict the weight loss of a particular subject who switches her diet.

The longitudinal weight loss study also involves several measurements on each of a relatively large number of independent subjects. If the study instead involved a large number of measurements on each of a small number of clusters, Bellamy et al. (2000, 2005) showed that linear mixed models with a penalized quasilikelihood (PQL) estimator performed well in estimating cluster-level covariate effects. The work of Bellamy et al. (2000, 2005) is especially useful for group randomized trials, which often involve a small number of large clusters.

1.2 Why Traditional Approaches for Independent Measurements Are Not Appropriate for Analysis of Longitudinal Weight Loss Study

If only one weight measurement was collected on each subject, or if each subject's experience was reduced to a single measurement, then standard approaches for analysis of independent measurements could be applied. For example, the mean weight change, evaluated by subtracting the baseline from final weight on each subject, could be compared between groups using the Student's t-test, or via linear regression. The proportion of subjects who had lost at least five percent of their body weight, as assessed at their final visit, could be compared using the Chi-Square test, or via logistic regression (Hilbe, 2009).

However, if more than two serial measurements were collected on each subject, as is the case in many longitudinal studies, standard statistical approaches that assume independence of measurements may not be appropriate. For example, in comparing the average change in weight (or in the likelihood of having lost at least five percent of body weight) over time between treatment groups, standard linear or logistic regression approaches do not adjust for the similarity, and resultant correlation, between measurements on the same subject.

Failure to adjust for intra-subject correlation could have negative consequences that include potential: (1) loss in efficiency in estimating the regression parameter that results in a decreased ability to detect a significant difference between treatment groups with respect to change in weight over time; (2) loss of useful information regarding the study outcome, for example within subject correlations that decay slowly over time could suggest that weights are relatively stable within subjects; and (3) loss of information that could have resulted in a cost savings for other investigators when planning their own studies, for example knowledge that the intra-subject correlations are likely to be larger could allow for a reduced sample size when planning a study that will compare change over time between weight loss interventions.

1.3   Attractive Features of Both QLS and GEE

In contrast to standard approaches that assume independence, GEE and QLS are *population-averaged* approaches that account for the correlation between repeated measurements on the same subject. They would therefore both be appropriate for analysis of the weight loss study considered in this section. In addition, they share an important advantage, namely they allow for straightforward implementation of the following standard steps in the analysis:

Specification of the Models: Specification of the models is straightforward, especially for those with prior experience with linear, logistic, or Poisson regression, because QLS and GEE first require specification of the usual generalized linear model that relates the expected value of the outcome variable with covariates measured on each subject. For example, for comparison of weight change over time between treatment groups, the usual linear regression model could be specified that regresses weight on baseline weight, time, and an interaction term that is constructed as the product of time and treatment group.

In addition, the usual logistic model could be specified to compare change in the likelihood that a subject has lost at least five percent of their body weight since baseline. The logistic model would involve a binary outcome that takes value 1 if a subject lost at least five percent of their body weight since baseline, and that takes value 0 otherwise; as for the linear model, covariates might include time, an indicator variable for group, and a time by treatment group interaction term. Of course, choice of the model is linked to the null and alternative hypotheses that are appropriate for our particular scientific questions of interest. In a thought-provoking article, Shaw and Proschan (2013) present elegant examples of statistical testing scenarios with hypotheses whose selection is not as straightforward as might appear upon first inspection.

After specification of the *regression model* for the expected value of the outcome, an additional step is then necessary to account for the potential intra-subject correlation of measurements: The analyst needs to select a *working correlation structure* that describes the pattern of association among the repeated measurements on each subject.

Specification of particular correlation structures will involve different assumptions about the pattern of association. For example, in the weight loss study we might anticipate that weights collected at weeks 1 and 2 post baseline on a subject will be more similar, and therefore more highly correlated, than weights collected at weeks 1 and 10 post baseline. The analyst might therefore select a working correlation structure that allows the correlation between measurements to depend on separation in time. If the measurements on subjects are unequally spaced in time, then a particular structure that is appropriate for unequally spaced data (see Chapter 5) might be applied. After specifying a structure that is thought to be most biologically plausible, additional structures might also be identified to allow for assessment of the sensitivity of results to the choice of working correlation structure. Several approaches (see Chapter 8) are also available for selection of a final working correlation structure.

GEE and QLS only require specification of models for the expected value of the outcome variable and correlation structure, but not for higher order moments, because both approaches are estimating equation procedures for semi-parametric models where the full likelihood is not specified.

Fitting the Models: Implementation of QLS and GEE is straightforward because both approaches can be readily implemented in most of the major statistical software packages. GEE can be applied using the **PROC GENMOD** procedure in SAS; the **geepack package** in R Statistical software (Yan, 2002); the **xtgee** procedure in Stata; the **genlin** procedure in SPSS version 15.0; and the **QLSPACK** procedure in MATLAB (Ratcliffe and Shults, 2008). QLS can be implemented in the **qlsinr** package in R (Xie and Shults, 2009); the **xtqls** procedure in Stata (Shults et al., 2007); the **QLSPACK** procedure in MATLAB (Ratcliffe and Shults, 2008); and the **QLS** macro in SAS (Kim and Shults, 2010).

Implementation of GEE or QLS in a particular software package will usually involve specifying a command that is very similar to the command for linear (or logistic or Poisson) regression, but modified to include specification of a working correlation structure, coupled with specification of the variables that indicate the subject identification numbers and the timings of measurements on each subject, or cluster. The modifications of the usual command for logistic or linear regression stem from the need to differentiate between subjects or clusters; order measurements within clusters; and identify a pattern for the correlations. Specification of additional options may also be important, for example, to choose between two types of estimated asymptotic covariance matrices for the estimated regression parameter. The relative ease with respect to specification of models for GEE and QLS therefore translates to corresponding ease with respect to implementation of these approaches in existing statistical software packages.

Interpretation of the Results: Interpretation of results is straightforward, especially for those with experience in fitting generalized linear models, because, in moving from standard regression approaches for independent measurements to QLS and GEE, the interpretation of the regression parameter remains virtually unchanged. For example, if the coefficient for the interaction term differs significantly from zero in the linear (or logistic) regression model for the weight loss study, this would indicate that the change in weight (or likelihood that a subject has lost at least five percent of their body weight since baseline) differs significantly between the two treatment groups over time.

In contrast to the regression parameters, which were directly involved in the primary analysis of the weight loss study, the correlation parameter is typically not involved in the primary hypotheses. Direct interpretation of the correlation parameters is therefore not always necessary. However, it would be beneficial to report the estimated values of the correlations, because they are necessary for sample size calculation of longitudinal studies, and therefore will be very important for other researchers who plan to conduct similar weight loss studies in the future.

Other situations in which the correlations are of direct interest include interventional trials, for which a misspecified pattern of association could have negative consequences. For example, Fitzmaurice (1995) showed that incorrectly assuming independence of outcomes that are highly correlated within clusters (or subjects) could result in a serious loss of efficiency in estimating the regression parameters associated with the covariates that vary within subjects. In the most extreme situation, incorrectly ignoring the correlation could result in poor precision in estimating the time by treatment group interaction term, that results in non-significant study results and subsequent dismissal of what might otherwise have been classified as a promising new treatment. In addition, the correlations may be informative in their own right. In general, Fitzmaurice (1995) recommends that "some attempt should generally be made to model the association between responses, even when the association is regarded as a nuisance characteristic of the data..."

1.4   When QLS Might Be Considered as an Alternative to GEE

Although GEE and QLS are both useful in analysis of correlated data, QLS does have some advantages over GEE that make it especially attractive for analysis of data from medical (and other types of) studies. Each of these advantages will be explored in subsequent chapters:

- In some situations QLS may estimate the regression parameter more efficiently than GEE; this will typically lead to smaller p-values, with subsequent improved likelihood of obtaining significant results for a clinical trial (or for another type of study).

- The QLS estimation procedure may converge for some datasets for which GEE has failed to converge. Therefore, QLS might be considered as an alternative approach when GEE fails. (However, it should be noted that failure to converge could be an indication that there is a serious problem with respect to the model assumptions. For example, failure to converge could be a result of specification of the incorrect correlation structure, so that the pattern of association was not correctly specified.)

- QLS allows for implementation of some potentially useful correlation structures that are not available in the major software packages that implement GEE. These structures are appropriate for analysis of multivariate longitudinal data, or more generally, for data with multiple sources of correlation; measurements that stabilize over time; longitudinal data with variances that change over time; and for data collected within families and for which the intra-familial correlations are expected to vary according to familial relationship. Implementation of these new correlation structures with QLS may prevent substantial loss in efficiency in estimating the regression parameter (Shults et al., 2004, 2006a; Xie et al., 2010).

Although QLS does have some advantages over GEE, it is important to note that it is not always necessary to choose *between* QLS and GEE. For example, in an analysis of data from a clinical trial, both methods could be applied, to assess the sensitivity of results to different model assumptions. Or, in a grant proposal for a medical study, the analysis plan might propose the application of both QLS and GEE, with QLS considered to be the "back-up approach" should GEE fail to converge, or the primary approach should QLS allow for implementation of a particular correlation structure that is biologically plausible but is not available for GEE. That QLS can be viewed as an enhancement to GEE is actually a benefit of the approach, because this improves its appeal to medical reviewers, many of whom are very familiar with and comfortable with GEE. Subsequent chapters will explain how to implement QLS in all phases of an analysis, from sample size computation, to selection of an appropriate working correlation structure, and assessment of the fit of competing models. In addition, we will describe alternate approaches that may offer advantages to both QLS and GEE in some situations; these will include the method of quadratic inferences functions (QIF) that was introduced by Qu et al. (2000) and was further explored and extended in Qu and Li (2006), Qu and Lindsay (2003), Qu and Song (2004).

1.5 Motivating Studies

This section presents several studies that will be referred to throughout the book. This is partly for convenience, so that the reader can easily refer back to a common location for study descriptions, and also to provide the reader with some additional description of the types of studies to which QLS might be applied. In addition, in the final chapter we present a worked example of data from an additional study of Bipolar II type depression.

1.5.1 Longitudinal Study of Obesity in Children Following Renal Transplant: With Binary and Continuous Measurements That Are Unequally Spaced in Time

In a longitudinal study conducted at the Children's Hospital of Philadelphia, body mass index (BMI) z-score and related variables were repeatedly measured on 100 children following a kidney transplant. Massive doses of steroids are usually given following transplant, so it was anticipated that BMI z-scores would increase and then perhaps taper off in the months following kidney transplant. The study participants were children, so that obesity was defined by comparing each child's BMI z-score with the expected z-score for a child of the same age and gender: Obesity was defined as having a BMI z-score greater than the 95th percentile for age and gender based on the 2000 National Center for Health Statistics (NCHS) growth curves and the recommendations of the American Academy of Pediatrics (O'Brien et al., 2004).

In order to facilitate sharing of data from this study, 10 percent of observations were dropped prior to saving the dataset random_small.dta that was also described by Shults et al. (2007) and is available at the following website:

http://www.cceb.upenn.edu/~sratclif/QLSproject.html.

The variables in random_small.dta are described in the following table:

Variable Name	Description
id	Subject identification number
month	Month of measurement
bmiz	BMI z-score
obese	Indicator variable for obesity

Important Features of the Data from This Study:

- *The inclusion of a binary outcome variable:* In addition to the continuous BMI z-score, this study included a binary outcome that indicated whether or not the child was obese at a particular measurement occasion. Methods appropriate for the analysis of binary measurements should therefore be implemented for analysis of the data from this trial.

- *Unequal temporal spacing of a planned common set of measurement occasions:* The gap in time between consecutive measurements on a subject ranged from 2 to 36 months, with a mean gap of 7.8 months. The study design planned for measurements to be collected at 1, 3, 6, 12, 24, and 36 months post baseline during the first 3 years after transplant. However, as noted above, 10 percent of the measurements were dropped from the dataset, in order to facilitate sharing of these data. In addition, some subjects missed their planned visits. Each patient therefore had measurement timings that were a subset of the common set of measurement times $\{1, 3, 6, 12, 24, 36\}$. Because BMI z-scores on a subject tend to be more similar when they are measured more closely together in time, it could be informative to implement a structure (see Chapter 5) that allows the correlation between measurements to depend on their separation in time. Implementation of an unstructured matrix is also possible and will be considered as well.

1.5.2 *Longitudinal Study of Sentence Recognition Scores That Stabilize over Time in a Hearing Recognition Study*

Gantz et al. (1988) compared five types of cochlear (implanted hearing aids) in patients who were profoundly, bilaterally deaf. We consider data discussed by Nunez-Anton and Woodworth (1994), who published a subset of the Gantz data that pertained to two of the implant types, which they referred to as types A and B. Implant types A and B were implanted in 23 and 21 patients, respectively. At 5 to 6 weeks after transplant, the patients were electronically connected to an external speech processor; they were then administered sentence recognition tests 1, 9, 18, and 30 months later. The data include the test scores for each patient, which represented the percentage of sentences that were recognized during each sentence recognition test. The variables in the dataset Nunez_Anton_Woodworth.raw from this trial are described in the following table:

Variable Name	Description
id	Subject identification number
month	Month of measurement
group	Is 1 for implant type A and is 0 for type B
score	Percentage sentences recognized
abovemedian	Is 1 if score is above the overall median

Important Features of the Data from This Study:

- *Unequal temporal spacing of measurements:* The study planned for hearing tests to be administered at 1, 9, 18, and 30 months after transplant and subsequent connection to an external speech processor. The design of the study therefore involved the collection of measurements that were unequally spaced in time; further imbalance was introduced when patients missed their tests intermittently, so that some patients had measurements missing between their first and final test. It is also important to note that, as in the previous example of renal transplant in children, patients had a common potential set of measurement times. Each patient's measurements times were a subset of $\{1, 9, 18, 30\}$.

  Common timings are practically more easy to deal with using a traditional GEE approach. For example, if all patients had complete data, then we might fit an unstructured correlation matrix with GEE, which would not force a particular choice of structure but would assume that the correlation between any two particular measurements on a subject (e.g., the 3rd and 4th measurements) is the same for all patients. We could then assess the values of the correlations, to see if the pattern in the correlations is consistent with a particular structure, for example, the AR(1) that we will discuss in Chapter 3. If measurement times are not common for all subjects, as was the case for this study, then application of an unstructured matrix is still possible for GEE. However, convergence can be problematic. In fact, as we shall see, implementation of an unstructured matrix for the data from this study will result in a failure to converge for GEE.

  Another approach is to fit GEE with an exchangeable structure that assumes a constant correlation between all measurement occasions. A drawback to assuming an exchangeable structure is that it might not be reasonable to assume that the correlation between measurements does not depend on their spacing in time, that is, that the correlation is the same, whether two measurements were collected at months 1 and 9 versus months 1 and 30. In Chapter 5 we will also discuss implementation of the Markov correlation structure with QLS; the Markov structure will allow the correlations between measurements to depend on their separation in time.

- *Stabilization in measurements over time:* Test scores were expected to stabilize over time for two reasons: First, patients were expected to gain familiarity with the test questions, which would result in a stabilization in their scores. In addition, test scores were expected to stabilize following an initial post-transplant improvement in hearing. Figure 1.1, which displays the mean test score versus month of measurement, suggests that test scores did indeed improve prior to leveling off following transplant. It may therefore be important to implement a correlation

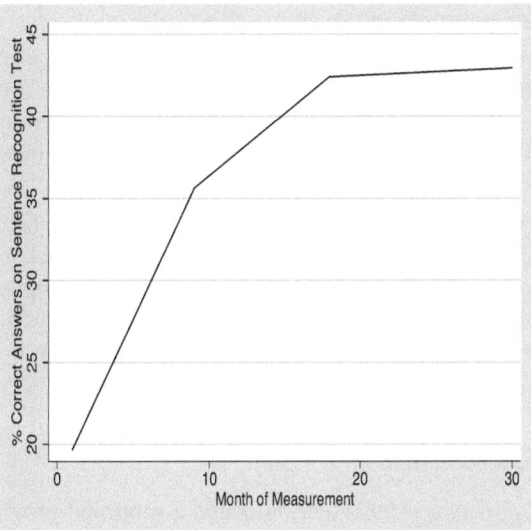

Figure 1.1 *Average of percentage recognition test score versus month of measurement.*

structure that is appropriate for measurements that stabilize over time. We will discuss such a structure, the *Generalized Markov*, in Chapter 5.

1.5.3 Longitudinal Study for Comparison of Two Treatments for Toenail Infection

De Backer et al. (1996) presented results from a randomized longitudinal study to compare two treatments for a common toenail infection. Data from this study were described in Table 2.3 of Molenberghs and Verbeke (2005) and they were also considered in Kim and Shults (2010). In the trial, 294 patients were randomized to one of two treatment groups (A and B). They were then measured at baseline, and again at 1, 2, 3, 6, 9, and 12 months post baseline, to determine the presence or absence of a severe toenail infection. The outcome variable is dichotomous and takes value y_{ij} = 1 if subject i has a severe toenail infection at measurement occasion j and takes value y_{ij} = 0 otherwise. Variables from the dataset *toenail.raw* are described in the following table:

Variable Name	Description
id	Patient identification number
time	Month of measurement
treatment	Is 1 for treatment A and is 0 for treatment B
interaction	Product of time and treatment

Important Features of the Data from This Study:

• *Unequal temporal spacing of measurements:* The study design planned for measurements to be collected at times that were common (the same for all subjects) and unequally spaced. Although it is unlikely that the actual data collection took place exactly when planned, we do not have access to the actual date of measurements; as a result, all subjects who did not miss visits were seen at 0, 1, 2, 3, 6, 9, and 12 months. However, some patients dropped out of the study and some missed visits intermittently (i.e., missed a visit but returned for a later measurement), so that each patient's measurements times were a subset of $\{0, 1, 2, 3, 6, 9, 12\}$.

- *Dichotomous outcome variable:* The outcome for analysis is binary and indicates whether or not the patient had a toenail infection at each measurement occasion. There is currently some controversy regarding the implementation of semi-parametric approaches such as GEE and QLS for the analysis of binary data. We will discuss the issues surrounding longitudinal binary data in Chapter 7 and will demonstrate QLS for their analysis for this trial.

1.5.4   Multivariate Longitudinal Dataset

Weissfeld and Kshirsagar (1992) demonstrated a modified growth curve model for a study in which patients in an intensive care unit were treated with three methods of suctioning their endotracheal (breathing) tubes. To assess the quality of the suctioning methods, oxygen saturations were measured on each patient at baseline, first suctioning pass, second suctioning pass, third suctioning pass, and 5 minutes after suctioning. Higher oxygen saturation levels for a particular suctioning method would suggest that the method was successful in clearing the artificial airways of these patients. The variables in the dataset

```
example_multlevel.dta
```

from this trial were obtained from Table 3.7 (p. 65) of Davis (2002) and are described in the following table:

Variable Name	Description
id	Subject identification number
time	Measurement occasion
type	Method of suction
o2	Oxygen saturation
family	Artificial family variable
high	1 if o2 > 96; 0 otherwise

Important Features of the Data from This Study:

- *The inclusion of both a continuous and a binary outcome variable:* The oxygen saturation variable *o2* is continuous, while the variable *high* that takes value 1 if oxygen saturation is above 96, is a binary outcome variable.

- *The continuous outcome variable is not normally distributed:* The oxygen saturation variable *o2* deviates from normality, as is suggested by the following stem and leaf plot.

```
Stem-and-leaf plot for o2
```

```
 7*  | 9
 8*  |
 8*  |
 8*  |
 8*  | 3
 8*  |
 8*  |
 8*  |
 8*  |
 8*  | 8
 8*  | 99
 9*  | 000000
 9*  | 1111111
 9*  | 222222222222222222222
 9*  | 3333333333333333333333333333
 9*  | 44444444444444444444444444444444444444444444
 9*  | 5555555555555555555555555555555555555555555555
 9*  | 666666666666666666666666666666666666666666666666666666
        6666666666666
 9*  | 7777777777777777777777777777777777777777777777777
 9*  | 8888888888888888888888888888888888888888888
 9*  | 999999999999999999999999999999999999999
10*  | 000000000000000000000000000000000
```

Therefore, methods for multivariate longitudinal data that assume normality might not be appropriate for the analysis of oxygen saturation levels.

- *Two sources of correlation:* In an analysis that compares the three methods of suctioning simultaneously, there will be two sources of correlation in the data, due to the fact that two measurements on a subject will tend to be more similar if they represent the same type of suctioning approach, or if they are collected on the same measurement occasion. We will implement QLS for a correlation structure that is appropriate for data with multiple sources of correlation in Chapter 6.

1.5.5 Familial Dataset

Magnus et al. (2001) explored the relationship between birth weights within families using information collected from mother–father–child trios who were born in Norway between 1967 and 1998. The investigators included only families with one child who survived at least the first month of life. The data from this study were extracted from the Medical Birth Registry of Norway and were analyzed by Professor Sophia Rabe-Hesketh in her presentation for the German Stata Users Group Meeting in Berlin, June 2010. The variables in the dataset familybirthwt.raw were based on data obtained from the website http://www.stata.com/meeting/germany10/abstracts.html (in the zip file for Professor Rabe-Hesketh's section on Biometrical modeling of twin and family data in Stata) and are described in the following table:

Variable Name	Description
family	Family identification number
type	1 if father; 2 if mother; 3 if infant
bwt	Birth-weight of subject (in grams)
lowbwt	1 if birth-weight of subject was less than 2500 grams; 0 otherwise
byr	Year of birth minus 1967
male	1 if subject is male; 0 otherwise
first	1 if subject was first-born; 0 otherwise
mothage	Mother's age (in years) at time of birth of subject
midage	1 if mother's age was between 20 and 35 at birth of subject; 0 otherwise
highage	1 if mother's age was greater than 35 at birth of subject; 0 otherwise
M	1 if subject is the mother; 0 otherwise
F	1 if the subject is the father; 0 otherwise
K	1 if the subject is the infant; 0 otherwise

Important Features of the Data from This Study:

- *The inclusion of both a continuous and a binary outcome variable:* The birth-weight variable *bwt* is continuous, while the variable *lowbwt* that takes value 1 if birth-weight is below 2,500 grams is binary.

- *Intra-familial correlations are of interest:* In this dataset, birth-weights are available on 1,000 families that each include a mother, father, and infant. In Chapter 4 we will relate the birth-weight of subjects with variables that include their mother's birth-weight and whether or not the subject is male, or was first-born. In this analysis it will be important to estimate the correlation in birth-weight between father and infant (father–infant correlation), mother and infant (mother–infant correlation), and mother and father (mother–father). For example, Magnus et al. (2001) reported that because the mother–father correlations were extremely low, this implied that the non-negligible father–child correlations were explained by genetic effects. In Chapter 4 we will analyze these data using a working correlation structure that allows the correlations to vary according to familial relationship.

1.6   Summary

QLS is a computational approach for estimation of the correlation parameters that is in the framework of the popular GEE approach. Both QLS and GEE are useful in collaborations because their models are easy to implement and interpret. Furthermore, software is available for application of each approach. We view QLS as an enhancement to GEE that will allow for improved analysis in some situations, for example, when we wish to implement a correlation structure that is only available for QLS. In this text we will describe how to implement QLS, with discussions that include sample size calculation using the approach of Preisser et al. (2003b, 2007); selection of a working correlation structure; assessment of the fit of the models; and comparison with popular competing approaches that will include GEE and QIF. We will also provide a thorough discussion of the theoretical underpinnings of QLS and discuss some ongoing areas of research that relate to QLS and GEE.

1.7   Exercises

Exercise 1.1 *Suppose we are analyzing data from a randomized clinical trial that compared weight loss over time in patients who were randomized to either the Atkins or the Slim-Fast diet. Suppose also that four measurements were collected on each of 30 patients per diet group at baseline, and at 6, 12, and 18 months post baseline.*

   a. *Describe a regression model that might be specified to compare weight change since baseline between the two groups.*

   b. *Suppose that we use traditional linear regression to fit this model, that is we ignore the fact that we have repeated measurements on the patients in this study. What are some potential implications on the results of the analysis?*

   c. *Is there a way to reduce the repeated measurements per subject down to one measurement per patient, in such a way that allows for a meaningful comparison of weight loss between the two diets? What is a potential drawback to this approach?*

Exercise 1.2 *When comparing methods for statistics journals, authors often like to present examples for which one of the methods being assessed fails, but for which the other method is successful. For example, they might present a dataset for which QLS converges but GEE fails to converge. Or they might present an example for which statistical significance is attained for one method but not for its competitor. In practice, for example when analyzing data from a randomized clinical trial, how should we react to such situations? For example, what should we do first if GEE fails to converge, but QLS converges?*

Exercise 1.3 *Suppose that you need to conduct an analysis of longitudinal data and that you will also involve some non-statistician medical colleagues in the analysis. Your colleagues do have some experience with regression and t-tests and the like. What are some advantages of using QLS or GEE in your analysis with your non-*

statistician colleagues, assuming that both approaches are appropriate? What are some potential advantages of QLS relative to GEE?

Exercise 1.4 *When dealing with longitudinal or clustered data, an important step in the analysis will involve specifying the pattern of association in the data. For example, in a study of birth-weight of rats within litters, it might be reasonable to assume that the pairwise association of birth-weights is the same for any two rats within a litter. However, if measurements are collected over time, then there might be a temporal aspect to the pairwise association of the repeated measurements within subjects. For example, in a longitudinal weight loss study, the association between measurements might be greater (due to greater similarity) if the weights are taken 1 day, as opposed to 1 year, apart in time. Without specifying a particular pattern, provide a general description of the pattern of association you might expect to find in each of the following studies: (In particular, would you anticipate a temporal aspect to the correlation and would you ever expect the correlations to be negative?)*

a. *A study in which income is collected at one measurement occasion within husband–wife pairs.*

b. *A study in which body weight is measured on members of several wolf packs during a winter season when food sources are limited. For simplicity, assume that the social structure of the pack is such that there is one alpha male that will eat first, followed by other members of the pack that are all equal in social status.*

c. *A study in which multiple measures of bone strength are collected at different sites on a patient (e.g., wrist, ankle, femur) at one particular measurement occasion.*

d. *A study in which multiple measures of bone strength are collected at different sites on a patient (e.g., wrist, ankle, femur) at one particular measurement occasion. Then the process is repeated again 6 and 12 months later.*

Chapter 2

Review of Generalized Linear Models and Generalized Estimating Equations

2.1 Background

Nelder and Wedderburn (1972) introduced the term "generalized linear models" (GLM) when they extended the scoring method for maximum likelihood estimation (ML) for normal data to exponential family models. This allowed for a unified approach for analysis that included linear, logistic, and Poisson regression, among others, as special cases. Wedderburn (1974) further generalized GLM by noting that key components of a GLM involved the first 2 moments of the distribution of the outcome variable, which could be used to define a *quasi-likelihood* that could be used to generalize the assumption of an exponential family for GLM.

The introduction of GLM was extremely important, because it provided a framework for analysis of outcome variables that could be either continuous or discrete. However, one limitation of GLM was that it was developed for independent measurements. As a consequence, GLM would typically not be appropriate for the analysis of longitudinal data from a clinical trial, or for data from any study for which an assumption of independence of measurements is unreasonable (e.g., longitudinal data from a clinical trial).

Liang and Zeger's introduction of GEE extended the application of GLM to correlated data, via specification of a GLM for the outcome variable, coupled with specification of a working correlation structure to describe the pattern of association among the repeated measurements on a subject, or multiple measurements within a cluster (Liang and Zeger, 1986). QLS can then be viewed as a computational approach for estimation of the correlation parameters that is in the framework of GEE. QLS only differs with respect to GEE in its estimation of the correlations, but is identical in its approach to estimation of the regression parameter.

In this chapter we provide a brief description of GLM and quasi-likelihood, followed by a slightly more detailed discussion of GEE. Our goal is to describe the methods that provide a foundation for QLS, in order to facilitate the reader's understanding of our later descriptions of QLS. Readers who desire more detailed descriptions of the history and implementation of GLM and GEE are encouraged to consult the textbooks by McCullagh and Nelder (1989), Hardin and Hilbe (2007), Dobson and Barnett (2008), Hardin and Hilbe (2012), and Ziegler (2011) and the manuscripts that they cite. An excellent text on modern methods for analysis of correlated data is also provided by Song (2007).

2.2 Generalized Linear Models

2.2.1 Linear Model

Before describing the GLM, it will be helpful to recall the classical linear model that was generalized by Nelder and Wedderburn (1972). Consider measurements y_i and associated vectors of explanatory variables, or covariates, $\boldsymbol{x}_i = (x_{i1}, \cdots, x_{ip})'$, for $i = 1, 2, \ldots, m$. For y_i that are continuous and independent, we could specify a linear model that involves a *systematic* and a *random* component:

The *systematic component* links the covariates $\boldsymbol{x}_i$ measured on each subject with the expected value $E(y_i) = \mu_i$ by expressing μ_i as a function of covariates via the relation $E(y_i) = \mu_i = \boldsymbol{x}_i'\boldsymbol{\beta}$, where $\boldsymbol{\beta}$ is a $p \times 1$ regression parameter and $\eta_i = \boldsymbol{x}_i'\boldsymbol{\beta}$ is referred to as the *linear predictor*.

The *random component* involves the assumption that the measurements y_i are independent and normally distributed, with expected value $E(y_i) = \mu_i$ and variance $Var(y_i) = \phi$.

The linear model expresses the expected values μ_i as a linear function of the regression parameter $\boldsymbol{\beta}$. An interesting feature of this model is that if the principle of least squares were applied for estimation of $\boldsymbol{\beta}$, by minimizing the sum of squared residuals $\sum_{i=1}^{m} (y_i - \boldsymbol{x}_i'\boldsymbol{\beta})^2$ with respect to $\boldsymbol{\beta}$, then this would yield the same estimate for $\boldsymbol{\beta}$ as would application of the principle of maximum likelihood (ML), under an assumption of independence and normality for the y_i. However, the assumption of normality that is made in the *random component* allows for construction of confidence intervals and tests that are based on the normal distribution; another interesting feature of the linear model is that these tests and confidence intervals will be based on the normal distribution for *finite samples*. In contrast, we shall see that the GEE and QLS estimates of $\boldsymbol{\beta}$ will be asymptotically normal, which means that the sample sizes will have to be sufficiently large for the estimates to be "reasonably normal." There are many excellent texts available that provide a comprehensive discussion of linear models and related topics, for example, by Christensen (1987), Christensen (1997), Rencher and Schaalje (2007), and Monahan (2008).

2.2.2 Generalized Linear Model

GLMs generalize the assumptions regarding the components of a linear model so that the systematic and random components of a GLM can be expressed as follows:

The *systematic component* relates the covariates $\boldsymbol{x}_i$ measured on each subject with the expected values $E(y_i) = \mu_i$ via the relation $E(y_i) = \mu_i = g^{-1}(\boldsymbol{x}_i'\boldsymbol{\beta})$, where $g^{-1}(\eta)$ is the inverse of the *link function* $g(\eta)$ that links the covariates measured on each subject with the expected value of the outcome variable.

The *random component* involves the assumption that the measurements y_i are

Table 2.1 *Characteristics of Three Common Exponential Distributions (2.1)*

	Normal	Poisson	Bernoulli (B(1,θ_i))
ϕ	$\phi = \sigma^2$	1	1
$b(\theta_i)$	$\theta_i^2/2$	$exp(\theta_i)$	$ln(1+exp(\theta_i))$
$c(y_i,\phi)$	$-1/2(y_i^2/\phi + ln(2\pi\phi))$	$-ln(y_i!)$	0

independent and distributed according to an exponential family

$$f(y_i, \theta_i, \phi) = exp\left\{[y_i\theta_i - b(\theta_i)]/\phi + c[y_i, \phi]\right\}, \tag{2.1}$$

where ϕ and θ_i are referred to as the *dispersion* and the *canonical* parameters, respectively. Let $b'(\theta_i)$ and $b''(\theta_i)$ refer to the first and second derivative, respectively, of $b(\theta_i)$ with respect to θ_i. As shown by Nelder and Wedderburn (1972), the expected value and variance of the y_i for Equation (2.1) can be expressed as $E(y_i) = \mu_i = b'(\theta_i)$ and as $Var(y_i) = b''(\theta_i)\phi$, respectively.

The GLM modified the systematic component of the linear model by generalizing the assumption of an *identity* link function, for which $\mu_i = x_i'\beta$, to a more general inverse link function, for which $\mu_i = g^{-1}(x_i'\beta)$. Furthermore, the GLM generalized the assumptions of independence and normality in the random component of the linear model, to an assumption of independence and an *exponential family* distribution for the y_i in the GLM.

Note also that for the exponential distribution, both the mean and variance of the y_i were expressed as functions of the canonical parameter θ_i in Equation (2.1); the variances can therefore be expressed as functions of the expected values for this distribution, via the *variance function* $h(\mu_i)$, so that $Var(y_i) = \phi h(\mu_i)$. Table 2.1 displays the parameters involved in the exponential distribution (Equation (2.1)) that yield the normal, Poisson, and Bernoulli (B(1,θ_i)) distributions. (Note also that non-link functions and other distributions could also readily be considered. However, to illustrate the concepts in our brief review of GLM in this chapter, we focus on the normal, Bernoulli, and Poisson distributions for measurements that are continuous, binary, or represent counts, respectively.)

For example, consider the Bernoulli (B(1,θ_i)) distribution

$$f(y_i, \theta_i) = exp(y_i\theta_i)/(1+exp(\theta_i)), \tag{2.2}$$

for $y_i \in \{0,1\}$, that can be expressed in the form of the exponential family distribution (2.1), for $b(\theta_i) = ln(1+exp(\theta_i))$; $\phi = 1$; and $c(y_i; \phi) = 0$. We can then obtain the mean and variance of the Bernoulli distribution via

$$\begin{aligned} \mu_i &= b'(\theta_i) \\ &= exp(\theta_i)/(1+exp(\theta_i)) \end{aligned}$$

and

$$
\begin{aligned}
Var(y_i) &= b''(\theta_i)\phi \\
&= exp(\theta_i)/(1 + exp(\theta_i)) - \\
&\quad (exp(\theta_i)/(1 + exp(\theta_i)))^2,
\end{aligned}
$$

respectively. Note that the variance function $\phi h(\mu_i)$, which expresses the variance as a function of the mean for this distribution, satisfies

$$
Var(y_i) = \phi h(\mu_i) = \mu_i(1 - \mu_i).
$$

In this text we will focus on *canonical link functions*, which are obtained by equating the linear predictor η_i with the canonical parameter θ_i. For example, if we specify the *canonical link* function for the Bernoulli distribution, so that $\theta_i = \eta_i = x_i'\beta$, then $\mu_i = exp(x_i'\beta)/(1 + exp(x_i'\beta))$, so that $x_i'\beta = ln(\mu_i/(1 - \mu_i))$, where $ln(\mu)$ is the natural log of μ. Next we summarize the canonical link and variance functions for the normal, Bernoulli, and Poisson distributions:

- *Normal distribution:* $E(y_i) = x_i'\beta$ and $Var(y_i) = \phi$, so that $g^{-1}(\gamma) = \gamma$ is the inverse identity link function, and $h(\mu) = 1$ is the variance function for the normal distribution.

- *Bernoulli distribution:* $E(y_i) = exp(x_i'\beta)/(1 + exp(x_i'\beta)) = \mu_i$ and $Var(y_i) = \mu_i(1 - \mu_i)$, so that $g^{-1}(\gamma) = exp(\gamma)/(1 + exp(\gamma))$ is the logistic inverse link function (with corresponding logit link function $g(\gamma) = ln(\gamma/(1 - \gamma))$), and $h(\mu) = \mu(1 - \mu)$ is the variance function for the Bernoulli distribution.

- *Poisson distribution:* $E(y_i) = exp(x_i'\beta) = \mu_i$ and $Var(y_i) = \mu_i$, so that $g^{-1}(\gamma) = exp(\gamma)$ is the inverse link function (with corresponding log link function $g(\gamma) = ln(\gamma)$) and $h(\mu) = \mu$ is the variance function for the Poisson distribution.

2.2.3   Estimation of the Parameters

The application of the principle of maximum likelihood for estimation of β in the exponential distribution (Equation (2.1)) when $\theta_i = x_i'\beta$ involves first obtaining the log-likelihood, which is the (natural) log of the likelihood function $\prod_{i=1}^{m} f(y_i, \theta_i, \phi)$. The log-likelihood has the following form:

$$
\sum_{i=1}^{m} \left(\frac{y_i\theta_i - b(\theta_i)}{\phi} + c(y_i, \phi) \right) \Bigg|_{\theta_i = x_i'\beta}. \tag{2.3}
$$

Next, to obtain the maximum likelihood estimate of β, we differentiate the log-likelihood (2.3) with respect to β, to obtain the following estimating equation for β:

$$
\sum_{i=1}^{m} x_i \frac{(y_i - b'(\theta_i))}{\phi} \Bigg|_{\theta_i = x_i'\beta} = 0, \tag{2.4}
$$

where x_i is the $p \times 1$ vector that was obtained by differentiating $\theta_i = x_i'\boldsymbol{\beta}$ with respect to $\boldsymbol{\beta}$ and $\mathbf{0}$ is a $p \times 1$ vector of zeros.

The estimating equation (2.4) is a nonlinear function of $\boldsymbol{\beta}$, so that an iterative approach such as iteratively re-weighted least squares can be applied to obtain a solution for $\boldsymbol{\beta}$; see Nelder and Wedderburn (1972) and Hardin and Hilbe (2007) for more details. Also see Hilbe (2009) for a comprehensive treatment of logistic regression.

To generalize Equation (2.4) to GEE in subsequent sections, it will be helpful to re-express this estimating equation as a function of the means μ_i and variances $Var(y_i)$ of the y_i. First, multiply the numerator and denominator of the functions in Equation (2.4) by $b''(\theta_i)$ to obtain

$$\sum_{i=1}^{m} b''(\theta_i) x_i \frac{(y_i - b'(\theta_i))}{b''(\theta_i)\phi}\bigg|_{\theta_i = x_i'\beta} = \mathbf{0}. \tag{2.5}$$

Next, recall that $b'(\theta_i) = \mu_i$ and $Var(y_i) = b''(\theta_i)\phi$; note also that $\frac{\partial \mu_i}{\partial \boldsymbol{\beta}} = b''(\theta_i)x_i$. Substitution into Equation (2.5) therefore yields the following equivalent GLM estimating equation for $\boldsymbol{\beta}$:

$$\sum_{i=1}^{m} \frac{\partial \mu_i}{\partial \boldsymbol{\beta}} Var^{-1}(y_i)(y_i - \mu_i) = \mathbf{0}. \tag{2.6}$$

2.2.4   Quasi-Likelihood

Note that the score equation (2.6) for ML estimation of $\boldsymbol{\beta}$ for GLM depends solely on the first 2 moments of the y_i. Wedderburn (1974) proposed relaxing the assumption of an exponential distribution for GLM. Rather than specify a particular form for the distribution of the y_i, he only assumed that the variance can be expressed as a particular function of the mean and perhaps a scalar parameter. In terms of the description provided in Section 2.2.2, quasi-likelihood estimation retains the assumption of the systematic component but discards the assumption of an exponential family distribution in the random component of the GLM. The motivation for quasi-likelihood stems from the observation that it is often more difficult in practice to determine the distribution than it is to identify the relationship between the mean and variance of the y_i.

Wedderburn (1974) defined a quasi-likelihood that depends solely on the mean and variance of the y_i and proved that the quasi-likelihood retains key features of the usual log-likelihood. He also proved that the quasi-likelihood reduces to the usual log-likelihood if and only if the data are from an exponential family, so that quasi-likelihood estimation includes GLM estimation as a special case. Godambe (2002, Chapter 9) gives a history of quasi-likelihood and describes its roots in two ideas of Gauss, namely the Gauss Markov theorem regarding least-squares (LS) estimation of $\boldsymbol{\beta}$ and the fact that the ML and LS estimates of $\boldsymbol{\beta}$ are uniquely identical for the normal distribution.

2.3   Generalized Estimating Equations

2.3.1   Notation for Correlated Data

To make the transition to correlated data, we now assume that *vectors* of measurements $Y_i = (y_{i1}, \cdots, y_{in_i})'$ were collected on each subject i, for $i = 1, 2, \ldots, m$. Individual measurements y_{ij} now involve two subscripts, i that indicates subject number, and j that indicates measurement number within subject. The notation from previous sections is then easily extended by replacing subscript i with the two subscripts ij. For example, covariates associated with y_{ij} are now denoted by $x_{ij} = (x_{ij1}, \cdots, x_{ijp})'$.

To understand the extension from GLM to GEE, it is helpful to re-express Equation (2.6) using the more general notation. Define the expected value of Y_i as $E(Y_i) = (\mu_{i1}, \mu_{i2}, \cdots, \mu_{in_i})' = \mu_i$. Next, let ϕA_i represent the diagonal matrix for which the jth diagonal element is the variance of y_{ij}, so that $\phi A_i = diag(var(y_{i1}), \cdots, var(y_{in_i})) = \phi diag(h(\mu_{i1}), \cdots, h(\mu_{in_i}))$. Also, let $D_i = \frac{\partial \mu_i}{\partial \beta} = \left(\frac{\partial \mu_{ij}}{\partial \beta_k}\right)$, so that element (j, k) of D_i is the derivative of μ_{ij} with respect to β_k. We can then re-express Equation (2.6) to obtain the following equivalent GLM estimating equation for β:

$$\sum_{i=1}^{m} D_i' A_i^{-1} (Y_i - \mu_i) = 0. \tag{2.7}$$

2.3.2   GEE Estimating Equation for β

Liang and Zeger (1986) extended GLM for measurements that are independent between subjects (or clusters) but are potentially correlated within subjects (or clusters). They first noted that the estimating equation (2.7) for a GLM can be expressed as a function of the correlation matrices for each subject: GLM assumes that the measurements within subjects are independent, so that the correlation matrix $Corr(Y_i)$ of measurements on subject i can be expressed as an $n_i \times n_i$ identity structure I_{n_i}. As a result, $A_i = A_i^{1/2} I_{n_i} A_i^{1/2} = A_i^{1/2} Corr(Y_i) A_i^{1/2}$ for independent measurements; substitution into Equation (2.7) then yields the following equivalent GLM estimating equation for β:

$$\sum_{i=1}^{m} D_i' A_i^{-1/2} Corr(Y_i)^{-1} A_i^{-1/2} (Y_i - \mu_i) = 0, \tag{2.8}$$

where $Corr(Y_i)$ is an $n_i \times n_i$ *identity matrix* that describes the pattern of association among the n_i *independent measurements* on subject i, for $i = 1, 2, \cdots, n_i$.

In order to extend GLM for correlated data, Liang and Zeger directly generalized the GLM estimating equation (2.8) that was obtained via application of the principle of maximum likelihood for independent measurements, for which $Corr(Y_i) = I_{n_i}$, to measurements that are correlated within subjects, for which $Corr(Y_i)$ is modeled with an $n_i \times n_i$ patterned correlation matrix $R_i(\alpha)$ that depends on correlation parameter α. They therefore relaxed the assumption of an $n_i \times n_i$ identity correlation matrix in (Equation (2.8)) to obtain the **GEE estimating equation for $\boldsymbol{\beta}$**:

$$\sum_{i=1}^{m} D_i' A_i^{-1/2} R_i^{-1}(\boldsymbol{\alpha}) A_i^{-1/2} (\boldsymbol{Y}_i - \boldsymbol{\mu}_i) = 0, \tag{2.9}$$

where $R_i(\boldsymbol{\alpha})$ is the working correlation structure, which may or may not equal $Corr(Y_i)$ that describes the pattern of association among the measurements y_{ij} on subject i. (Particular structures and the concept of *working structure* will be described in greater detail in the next two sections.) Note also that the GEE estimating equation (2.9) involves a particular parametrization of the covariance matrix of $\boldsymbol{Y}_i$, because $Cov(\boldsymbol{Y}_i) = \phi A_i^{1/2} Corr(Y_i) A_i^{1/2}$, so that $Cov^{-1}(\boldsymbol{Y}_i) = 1/\phi A_i^{-1/2} Corr^{-1}(Y_i) A_i^{-1/2}$.

The GEE estimating equation (2.9) can also be expressed as a function of the *Pearson residuals* $z_{ij} = (y_{ij} - \mu_{ij})/\sqrt{h(\mu_{ij})}$:

$$\sum_{i=1}^{m} D_i' A_i^{-1/2} R_i^{-1}(\alpha) \boldsymbol{Z}_i = 0, \tag{2.10}$$

where $\boldsymbol{Z}_i = A_i^{-1/2} (\boldsymbol{Y}_i - \boldsymbol{\mu}_i)$ is the $n_i \times 1$ vector of Pearson residuals z_{ij} on subject i, for $i = 1, 2, \cdots, m$.

The GEE in Equation (2.9) is therefore truly a *generalized* estimating equation because it was obtained by directly incorporating correlation parameters into the original GLM estimating equation for $\boldsymbol{\beta}$.

2.3.3 Working Correlation Structures Available for GEE

Liang and Zeger (1986) suggested several structured correlation matrices $R_i(\alpha)$ that could be used to describe the pattern of association among the repeated measurements on the ith subject. Here we define the structured matrices and provide the *feasible interval* for each structure (Chaganty and Shults, 1999), which is the interval on which α will yield a positive-definite correlation matrix.

- *Exchangeable:* For this structure (also known as equicorrelated), all pairwise intra-cluster correlations are equal, so that $Corr(y_{ij}, y_{ik}) = \alpha$. An $n_i \times n_i$ exchangeable structure has the following form:

$$\begin{pmatrix} 1 & \alpha & \cdots & \alpha \\ \alpha & 1 & \cdots & \alpha \\ \vdots & \vdots & \ddots & \vdots \\ \alpha & \alpha & \cdots & 1 \end{pmatrix}_{n_i \times n_i}. \tag{2.11}$$

The feasible interval for this structure is $(-1/(n_{max}-1), 1)$, where n_{max} represents the maximum value of n_i over $i = 1, 2, \ldots, m$. The exchangeable structure might be plausible for cross-sectional studies. For example, in an analysis that measures weights of rat pups at birth, it might be reasonable to assume that the weights of rat pups are similar within litters and are therefore correlated, but that no two rat pups are any more (or less) similar than any other two rat pups.

- *First-order autoregressive AR(1):* This structure assumes that the correlation between repeated measurements on a subject depends on their separation in order of measurement, so that $Corr(y_{ij}, y_{ik}) = \alpha^{|j-k|}$. An $n_i \times n_i$ AR(1) structure has the following form:

$$
\begin{pmatrix}
1 & \alpha & \alpha^2 & \dots & \alpha^{n_i-1} \\
\alpha & 1 & \alpha & \dots & \alpha^{n_i-2} \\
\alpha^2 & \alpha & 1 & \dots & \alpha^{n_i-3} \\
\vdots & \vdots & \vdots & \ddots & \vdots \\
\alpha^{n_i-1} & \alpha^{n_i-2} & \alpha^{n_i-3} & \dots & 1
\end{pmatrix}_{n_i \times n_i} . \tag{2.12}
$$

The feasible interval for this structure is $(-1, 1)$. The AR(1) structure forces the correlation between two measurements on a subject to decline (in absolute value) with increasing separation in measurement occasion and is perhaps most plausible for longitudinal studies in which the measurements are equally spaced in time. For example, if $\alpha = 0.90$, then the correlation between any consecutive two measurements on a subject is 0.90, while the correlation between measurements that are separated by two measurement occasions is smaller and is given by $0.90^2 = 0.81$. We will also see in Chapter 7 that in some situations, this structure is a natural structure for the analysis of correlated binary measurements.

- *Tri-diagonal:* This structure is also referred to as stationary 1-dependent. It assumes that the correlation between measurements on a subject is constant for measurements that are separated by one measurement occasion, so that $Corr(y_{ij}, y_{ik}) = \alpha$ for $|j-k| = 1$, and is zero otherwise. This structure was implemented in Liang and Zeger (1986) and in the standard software packages that implement GEE. An $n_i \times n_i$ tri-diagonal structure has the following form:

$$
\begin{pmatrix}
1 & \alpha & 0 & \dots & 0 \\
\alpha & 1 & \alpha & \dots & 0 \\
0 & \alpha & 1 & \dots & 0 \\
0 & \vdots & \vdots & \ddots & \vdots \\
0 & 0 & 0 & \dots & 1
\end{pmatrix}_{n_i \times n_i} . \tag{2.13}
$$

The feasible interval for this structure is $(-1/c_{max}, 1/c_{max})$, where $c_{max} = 2\sin\left(\frac{\pi[n_{max}-1]}{2[n_{max}+1]}\right)$ and n_{max} is the maximum value of n_i over $i = 1, 2, \dots, m$; this interval is approximately $(-1/2, 1/2)$ for large n and contains $(-1/2, 1/2)$ for all n.

- *Unstructured:* This structure does not assume any pattern for the intra-subject correlations, so that the assumed correlation between Y_{ij} and Y_{ik} is α_{jk}. An $n_i \times$

n_i unstructured matrix has the following form:

$$
\begin{pmatrix}
1 & \alpha_{12} & \alpha_{13} & \cdots & \alpha_{1n_i-1} \\
\alpha_{12} & 1 & \alpha_{23} & \cdots & \alpha_{2n_i-2} \\
\alpha_{13} & \alpha_{23} & 1 & \cdots & \alpha_{3n_i-3} \\
\vdots & \vdots & \vdots & \ddots & \vdots \\
\alpha_{1n_i-1} & \alpha_{2n_i-2} & \alpha_{3n_i-3} & \cdots & 1
\end{pmatrix}_{n_i \times n_i} . \tag{2.14}
$$

Although this structure is flexible, it is not parsimonious. For example, an $n_i \times n_i$ unstructured matrix will involve $n_i(n_i - 1)/2$ correlation parameters. One serious drawback in practice is that the iterative GEE estimation procedure will often fail to converge for the unstructured matrix. However, it can be a useful structure, especially for studies that involve a larger number of subjects; a relatively small number of measurements per subject; and a fixed set of outcomes or measurement occasions per subject, for which it is reasonable to assume that the correlation between the jth and k^{th} measurements is the same for all subjects.

- *Working Independent:* The identity matrix is straightforward to apply because β can then be estimated by solving the standard GLM estimating equation (2.7). An $n_i \times n_i$ identity structure has the following form:

$$
\begin{pmatrix}
1 & 0 & 0 & \cdots & 0 \\
0 & 1 & 0 & \cdots & 0 \\
0 & 0 & 1 & \cdots & 0 \\
0 & \vdots & \vdots & \ddots & \vdots \\
0 & 0 & 0 & \cdots & 1
\end{pmatrix}_{n_i \times n_i} . \tag{2.15}
$$

However, incorrect application of the working independence structure can result in a serious loss in efficiency in estimation of β, as shown by Wang and Carey (2003), Shults et al. (2006a), Sutradhar and Das (1999), Sutradhar and Das (2000), and many others.

In practice, if the number of measurements per subject is fairly small and if the measurement times are reasonably constant between subjects, a reasonable first step in the analysis might be to fit an unstructured correlation matrix with GEE. If the correlations are similar, we might then fit an equicorrelated structure; while if the correlations decline over time, we might fit an AR(1) correlation structure. Replacing the unstructured matrix with either the AR(1), tri-diagonal, or exchangeable structures would greatly improve the parsimony of the model for the correlation; the number of correlation parameters would be reduced to 1 because the latter structures depend on a single parameter α. In Chapter 8 we will compare different approaches for selection of a working correlation structure, including methods discussed by Barnett et al. (2010) and Shults et al. (2009).

We also note that the correlation structures described above are all subscripted by i, that is, we refer to the correlation structure for measurements on subject i as $R_i(\alpha)$ because we allow the dimensions of the structures to vary between subjects. Some authors do require an assumption of *balanced data*, for which the number of measurements per subject is n for all subjects; when the data are balanced, the correlation structures are $n \times n$ and any of the structures described above could be referred to as $R(\alpha)$, because the pattern of association would be identical for all subjects when the data are balanced. However, when possible we prefer to avoid the restrictive assumption of balanced data. With the exception of the unstructured matrix, we allow the number of measurements per subject (n_i) to vary between subjects; as a result, we refer to the correlation structures as $R_i(\alpha)$, to reflect that they vary in dimension between subjects. In addition, we will see that for some structures (e.g. the Markov structure that we will discuss in chapter 5) the pattern of association is not the same for all subjects, so that the association between the 2nd and 3rd measurements, for e.g., is not the same for all subjects.

2.3.4 The Concept of the Working versus the True Correlation Structure

In the previous section we described several structured matrices $R_i(\alpha)$ that could be used to model the pattern of association among repeated measurements on the ith subject (or cluster). It has been the tradition in the GEE literature to refer to the specified structure $R_i(\alpha)$ as a "working" correlation structure, to stress that the true structure is unknown and may have been misspecified. For example, an exchangeable form may have been selected for $R_i(\alpha)$; this choice will be wrong unless all pairwise correlations on the subjects (or clusters) are equal. Related to the concept of "working" structure is the view that the correlation parameters are nuisance parameters for GEE. For example, in a GEE analysis, typically tests will be conducted and confidence intervals will be constructed for the regression parameter only. This is in contrast to some other methods based on GEE that will be briefly introduced in Section 3.5, that do not treat the correlations as a nuisance.

In order to distinguish between correct versus incorrect specification of the pattern of association among the measurements within the ith subject or cluster, throughout this text we will refer to the structure that has been implemented in the analysis as the "working structure" $R_i(\alpha)$ that depends on "working" parameter α versus the true structure $T_i(\rho)$ that depends on the "true" parameter ρ. Unless otherwise specified, we will assume that the true structure has been correctly specified, so that $R_i(\alpha) = T_i(\rho)$ and $\alpha = \rho$. However, we will also periodically consider the impact of incorrect specification of the underlying correlation structure. For example, we shall see that GEE is in some ways robust to misidentification of the underlying correlation structure, but that there can be a price to pay when the underlying correlation structure has been misspecified.

2.3.5 Moment Estimates of the Dispersion and the Correlation Parameters

GEE employs moment estimates for the correlation parameters that are based on the *Pearson residuals* $z_{ij} = \frac{y_{ij} - \mu_{ij}}{\sqrt{h(\mu_{ij})}}$, where the variance of y_{ij} is $Var(y_{ij}) = \phi h(\mu_{ij})$. Moment estimates of α can be based on averages of products of the Pearson residuals because the expected value of their products is a function of the correlations and dispersion parameter ϕ:

$$
\begin{aligned}
E(z_{ij}z_{ik}) &= \phi\frac{E\left((y_{ij}-\mu_{ij})(y_{ik}-\mu_{ik})\right)}{\sqrt{\phi h(\mu_{ij})}\sqrt{\phi h(\mu_{ik})}} \\
&= \phi\frac{Cov(y_{ij},y_{ik})}{\sqrt{var(y_{ij})var(y_{ik})}} \\
&= \phi Corr(y_{ij},y_{ik}),
\end{aligned}
$$

where $Corr(y_{ij},y_{ik})$ is modeled by $r_{ijk}(\alpha)$, where $r_{ijk}=z_{ij}z_{ik}$. In addition, a moment estimate of the dispersion parameter ϕ can be based on an average of the sum of squared Pearson residuals because

$$
E(z_{ij}^{2}) = \phi.
$$

One general potential limitation to the use of moment estimates for α in the framework of GEE is that, in contrast to QLS, this approach does not yield only one form of an estimate for a particular correlation structure. For example, for the AR(1) correlation structure, the approach discussed in Liang and Zeger (1986, Example 4) could be applied. Because $E(z_{ij}z_{ik})=\phi\alpha^{|j-k|}$ for the AR(1) structure, substitution of the current estimates of z_{ij} and z_{ik} and taking natural logarithms yields the following equation:

$$
ln(\widehat{z}_{ij}\widehat{z}_{ik}) \approx ln(\phi)+|j-k|ln(\alpha). \tag{2.16}
$$

Liang and Zeger therefore suggested estimating $ln(\alpha)$ by the slope of the regression of $ln(\widehat{z}_{ij}\widehat{z}_{ik})$ on $|j-k|$. However, a serious drawback to this ad-hoc approach was that it did not guarantee estimates within the interval $(-1,1)$, which as noted earlier is the interval on which α yields a positive definite correlation matrix for the AR(1) structure. Shults and Chaganty (1998) noted that approximately 10 percent of simulation runs for the true AR(1) structure resulted in estimates of α that exceeded 1 in value. The weakness of this suggested regression approach for estimation of α is perhaps reflected in the fact that it has not been implemented in any of the standard software packages that implement GEE. In fact, the software packages SAS, Stata, and R all implement slightly different estimates of α for the AR(1) structure.

In this chapter we provide the GEE estimates of α that are implemented both in PROC GENMOD in SAS and in the GEEQBOX program for implementation of GEE in MATLAB (Ratcliffe and Shults, 2008).

For the *equicorrelated structure*, the GEE moment estimate is given by

$$
\widehat{\alpha}_{GEE-EQUI} = \frac{\sum_{i=1}^{m}\sum_{j\neq k}\widehat{z}_{ij}\widehat{z}_{ik}}{(N^{*}-p)\widehat{\phi}_{GEE}}, \tag{2.17}
$$

where

$$
N^{*} = \sum_{i=1}^{m}n_{i}(n_{i}-1);
$$

$$\widehat{\phi}_{GEE} = \frac{\sum_{i=1}^{m} \sum_{j=1}^{n_i} \widehat{z}_{ij}^2}{N - p}; \tag{2.18}$$

$N = \sum_{i=1}^{m} n_i$; $\widehat{z}_{ij}$ is the current estimate of the Pearson residual for subject i at time t_{ij}; and p is the dimension of $\boldsymbol{\beta}$.

For the *AR(1) and tri-diagonal structures*, the GEE moment estimate is

$$\widehat{\alpha}_{GEE-TRI} = \widehat{\alpha}_{GEE-AR1} = \frac{\sum_{i=1}^{m} \sum_{j=2}^{n_i} \widehat{z}_{ij} \widehat{z}_{i,j-1}}{(N^{**} - p)\widehat{\phi}_{GEE}}, \tag{2.19}$$

where $N^{**} = \sum_{i=1}^{m}(n_i - 1)$.

For the *unstructured correlation matrix*, GEE implements the following moment estimate for element j, k ($j \neq k$) of the matrix:

$$\widehat{R}_{GEE}[j,k] = \frac{\sum_{i=1}^{m} \widehat{z}_{ij} \widehat{z}_{ik}}{(m - p)\widehat{\phi}_{GEE}}. \tag{2.20}$$

The estimator (2.20) can be implemented for data that are missing intermittently, so that some subjects miss a visit prior to their last visit, in a study with measurement times that were planned to be the same for all subjects. For example, **PROC GENMOD** allows for specification of a variable that indicates the time of measurement. It then implements an "all-pairs" approach that only utilizes information on subjects with measurements at times t_j and t_k when estimating the correlation between y_{ij} and y_{ik}. However, greater imbalance in the data might result in greater instability and failure to converge.

2.3.6   Algorithm for Estimation

In order to implement GEE in an analysis, we first need to specify a generalized linear model to describe the mean and variance of y_{ij}: $E(y_{ij}) = g^{-1}(\boldsymbol{x}_{ij}'\boldsymbol{\beta})$ and $Var(y_{ij}) = \phi h(\mu_{ij})$. Next, we need to choose a working correlation structure $R_i(\alpha)$ to describe the pattern of association among the repeated measurements on subject i. The following algorithm can then be implemented for estimation of the parameters with GEE.

Algorithm for Estimation of the Regression and Correlation Parameters with GEE: First, select a starting value for the correlation parameters. Convenient starting values for α correspond to specification of an identify matrix, which is a special case of every correlation structure. Therefore, we could specify $\widehat{\alpha}_{GEE} = 0$. Next, alternate between Steps A and B (below) until there is convergence in the estimates, which is achieved when the current values of the estimates are considered sufficiently close to the previous values:

Step A : Update the regression parameter $\boldsymbol{\beta}$ via solution of the GEE estimating equation for $\boldsymbol{\beta}$:

$$\sum_{i=1}^{m} D_i' A_i^{-1/2} R_i^{-1}(\widehat{\alpha}_{GEE}) A_i^{-1/2} (\boldsymbol{Y}_i - \boldsymbol{\mu}_i(\boldsymbol{\beta})) = 0 \tag{2.21}$$

and

Step B : Update the estimate of α using a moment estimate that corresponds to the specified working correlation structure $R_i(\alpha)$. Particular choices for $\widehat{\alpha}_{GEE}$ were provided in Section 2.3.5.

Note that the GEE estimating equation (2.21) is nonlinear in $\boldsymbol{\beta}$. A modified Fisher-scoring method can be used in Step A to obtain an updated estimate $\widehat{\boldsymbol{\beta}}$ of $\boldsymbol{\beta}$ that is based on the previous estimates $\widetilde{\boldsymbol{\beta}}$ and $\widetilde{\alpha}$ as

$$\widehat{\boldsymbol{\beta}} = \widetilde{\boldsymbol{\beta}} + \left(\sum_{i=1}^{m} \widetilde{D}_i' \widetilde{V}_i^{-1} \widetilde{D}_i \right)^{-1} \left(\sum_{i=1}^{m} \widetilde{D}_i' \widetilde{V}_i^{-1} (\boldsymbol{Y}_i - \widetilde{\boldsymbol{\mu}}_i) \right), \qquad (2.22)$$

where $V_i = A_i^{1/2} R_i(\alpha) A_i^{1/2}$, and $\widetilde{A}_i$, $\widetilde{D}_i$, $\widetilde{\boldsymbol{\mu}}_i$ are A_i, D_i, $\boldsymbol{\mu}_i$ evaluated at $\widetilde{\boldsymbol{\beta}}$. The estimation procedure ends when $\widehat{\boldsymbol{\beta}} \approx \widetilde{\boldsymbol{\beta}}$.

In practice, however, it should not be necessary to program GEE because, as mentioned earlier, this approach is now available in the major statistical software packages. For example, GEE can be applied using the **PROC GENMOD** procedure in SAS; the **geepack** package in R Statistical software (Yan, 2002); the **xtgee** command in Stata; the **genlin** procedure in SPSS version 15.0; and the **QLSPACK** procedure in MATLAB (Ratcliffe and Shults, 2008).

Several diagnostics and checks for goodness of fit for GEE have been proposed in the literature. For example, Preisser and Qaqish (1996); Hammill and Preisser (2006); and Preisser et al. (2008) have developed deletion diagnostics for GEE that are currently available in SAS Version 9.2, in the **PROC GENMOD** procedure; these diagnostics measure the influence of a subset of observations on the estimated regression parameters and on the estimated values of the linear predictor.

2.3.7 *Asymptotic Distribution of the GEE Estimators and Estimates of Covariance*

Under standard regularity conditions and assuming that $\sqrt{m}$ consistent estimators are used for α and ϕ, Liang and Zeger (1986) proved that $\sqrt{m}(\widehat{\boldsymbol{\beta}} - \boldsymbol{\beta})$ is asymptotically Gaussian with mean 0 and covariance matrix $V_{\boldsymbol{\beta}}$ given by

$$V_{\boldsymbol{\beta}} = \lim_{m \to +\infty} m \left(\sum_{i=1}^{m} D_i' V_i^{-1} D_i \right)^{-1} \left[\sum_{i=1}^{m} D_i' V_i^{-1} Cov(\boldsymbol{Y}_i) V_i^{-1} D_i \right] \left(\sum_{i=1}^{m} D_i' V_i^{-1} D_i \right)^{-1},$$
$$(2.23)$$

where $Cov(\boldsymbol{Y}_i)$ is the true covariance matrix of $\boldsymbol{Y}_i$.

A "sandwich" estimator $\widehat{\Sigma}_S$ of $Cov(\widehat{\boldsymbol{\beta}})$ based on Equation (2.23) can then be obtained as

$$\widehat{\Sigma}_S = \left(\sum_{i=1}^{m} \widehat{D}_i' \widehat{V}_i^{-1} \widehat{D}_i \right)^{-1} \left[\sum_{i=1}^{m} \widehat{D}_i' \widehat{V}_i^{-1} \widehat{Cov}(\boldsymbol{Y}_i) \widehat{V}_i^{-1} \widehat{D}_i \right] \left(\sum_{i=1}^{m} \widehat{D}_i' \widehat{V}_i^{-1} \widehat{D}_i \right)^{-1}, \quad (2.24)$$

where $\widehat{Cov}(\boldsymbol{Y}_i) = (\boldsymbol{Y}_i - \widehat{\boldsymbol{\mu}}_i)(\boldsymbol{Y}_i - \widehat{\boldsymbol{\mu}}_i)'$ and $\widehat{D}_i$, $\widehat{\boldsymbol{\mu}}_i$ are D_i, $\boldsymbol{\mu}_i$ evaluated at $(\widehat{\alpha}, \widehat{\boldsymbol{\beta}})$.

If the correlation structure of Y_i has been correctly specified, we can estimate $Cov(Y_i)$ by $\widehat{\phi}_{GEE}\widehat{V}_i$, so that another estimator based on Equation (2.24) simplifies as the following "model" based estimator

$$\widehat{\Sigma}_M = \widehat{\phi}_{GEE}\left(\sum_{i=1}^{m}\widehat{D}_i'\widehat{V}_i^{-1}\widehat{D}_i\right)^{-1}, \tag{2.25}$$

where $\widehat{\phi}_{GEE}$ was defined in Equation (2.18).

The model-based estimator $\widehat{\Sigma}_M$ is a consistent estimator of $Cov(\widehat{\beta})$ only if the correlation structure has been correctly specified, while the sandwich estimator $\widehat{\Sigma}_S$ is generally viewed as being more robust to incorrect specification of the true covariance structure. Sandwich estimators of covariance date back to Huber (1967) and White (1982) and have been used to provide a consistent estimator of the covariance matrix, even when the parametric model fails to hold. For example, the sandwich estimator has been used to obtain a consistent estimate of the covariance when the constant variance assumption is violated in linear regression. However, Kauermann and Carroll (2001) demonstrated that the consistency of the sandwich estimator comes at a price in terms of increased variability and resultant undercoverage of confidence intervals, even for a simple linear regression model that was correctly specified.

Kauermann and Carroll (2001) also considered sandwich estimators for generalized linear models and GEE and noted that "the loss of efficiency of the sandwich variance estimate in nonnormal models differs from and can be worse than that occurring for normal models." For example, suppose that we are interested in linear combinations $z'\beta$ of the regression parameters. Kauermann and Carroll (2001) showed that the variances based on the sandwich estimator $\widehat{\Sigma}_S$ will be biased downward, so that the variances

$$\begin{aligned}
\widehat{Var(z'\beta)} &= z'\widehat{Cov(\widehat{\beta})}z & (2.26) \\
&= z'\widehat{\Sigma}_S z & (2.27) \\
&= \widehat{v}_s, & (2.28)
\end{aligned}$$

will tend to be too small. As a result, confidence intervals based on the traditional sandwich estimator $\widehat{\Sigma}_S$ will tend to be too narrow and the type I error rate of Wald tests may be inflated, especially for smaller sample sizes (e.g., ≤ 40 subjects). However, Kauermann and Carroll (2001) showed that undercoverage may not be an issue when the subjects have complete data and the covariance design is the same for all subjects, for example, in a moderately sized clinical trial in which all patients have n measurements and the covariates on each patient included a constant and scaled and centered timings of measurements, as in Kauermann and Carroll (2001, example 8). They therefore suggested revising the earlier suggestion of Diggle et al. (1994, p. 77) that the sandwich estimate should be used with care if the data come from a small number of experimental units, to include the additional condition that the covariate design should also differ between the units; thus, the sandwich covariance matrix should be used with caution for small samples and when the covariance design differs between subjects.

Recent research has focused on making a correction to the sandwich estimator $\widehat{\Sigma}_S$ so that it has better properties, especially for smaller samples; this includes a suggested adjustment by Kauermann and Carroll (2001). Following the discussions by Teerenstra et al. (2010), several estimators of $Cov(\widehat{\boldsymbol{\beta}})$ based on Equation (2.23) can be obtained using

$$\widehat{\Sigma}_T = \widehat{\Sigma}_1^{-1} \left[\sum_{i=1}^{m} \widehat{D}_i' \widehat{V}_i^{-1} \widehat{B}_i \widehat{Cov}(\boldsymbol{Y}_i) \widehat{B}_i \widehat{V}_i^{-1} \widehat{D}_i \right] \widehat{\Sigma}_1^{-1}, \tag{2.29}$$

where $\widehat{D}_i$, $\widehat{\mu}_i$ are D_i, μ_i evaluated at $(\widehat{\alpha}, \widehat{\boldsymbol{\beta}})$;

$$\widehat{\Sigma}_1^{-1} = \left(\sum_{i=1}^{m} \widehat{D}_i' \widehat{V}_i^{-1} \widehat{D}_i \right)^{-1}; \tag{2.30}$$

and each of the different estimators can be obtained by changing the value of the matrices $\widehat{B}_i$ for each estimator.

We first recover the model- and sandwich-based covariance estimates provided above. Start by assuming that the correlation structure of $\boldsymbol{Y}_i$ has been correctly specified, so that we can estimate $Cov(\boldsymbol{Y}_i)$ with $\widehat{\phi}_{GEE}\widehat{V}_i$. If we then replace $\widehat{Cov}(\boldsymbol{Y}_i)$ in Equation (2.29) with $\widehat{\phi}_{GEE}\widehat{V}_i$ and also assume that the matrices $\widehat{B}_i$ are identity matrices, so that $\widehat{B}_i = I_i$, then $\widehat{\Sigma}_T$ simplifies as the model based estimator that was provided in Equation (2.25)

$$\begin{aligned} \widehat{\Sigma}_M &= \widehat{\phi}_{GEE}\widehat{\Sigma}_1^{-1}\widehat{\Sigma}_1\widehat{\Sigma}_1^{-1} & (2.31) \\ &= \widehat{\phi}_{GEE}\widehat{\Sigma}_1^{-1}. & (2.32) \end{aligned}$$

Next, if we do not assume that the structure has necessarily been correctly specified, we can estimate $Cov(\boldsymbol{Y}_i)$ with $(\boldsymbol{Y}_i - \widehat{\boldsymbol{\mu}}_i)(\boldsymbol{Y}_i - \widehat{\boldsymbol{\mu}}_i)'$. If we then replace $\widehat{Cov}(\boldsymbol{Y}_i)$ in Equation (2.29) with $(\boldsymbol{Y}_i - \widehat{\boldsymbol{\mu}}_i)(\boldsymbol{Y}_i - \widehat{\boldsymbol{\mu}}_i)'$ and again set $\widehat{B}_i = I_i$, then $\widehat{\Sigma}_T$ simplifies as the sandwich estimator provided in Equation (2.24).

Next we use Equation (2.29) to obtain several corrections to Liang and Zeger's sandwich estimator that was provided in Equation (2.23). If we replace $\widehat{Cov}(\boldsymbol{Y}_i)$ in (2.29) with $(\boldsymbol{Y}_i - \widehat{\boldsymbol{\mu}}_i)(\boldsymbol{Y}_i - \widehat{\boldsymbol{\mu}}_i)'$ and, as in Lu et al. (2007), define the ith cluster leverage as

$$\widehat{H}_i = \widehat{D}_i \widehat{\Sigma}_1^{-1} \widehat{D}_i' \widehat{V}_i^{-1}, \tag{2.33}$$

then replacing $\widehat{B}_i$ with $\widehat{B}_i = \left(I_i - \widehat{H}_i \right)^{-1}$ in Equation (2.29) yields a bias corrected estimator proposed by Mancl and DeRouen (2001), which we will refer to as $\widehat{\Sigma}_{SMR}$; replacing $\widehat{B}_i$ with $\widehat{B}_i = \left(I_i - \widehat{H}_i \right)^{-1/2}$ in Equation (2.29) yields the bias-corrected estimator proposed by Kauermann and Carroll (2001), which we will refer to as $\widehat{\Sigma}_{SKC}$. An interesting point about $\widehat{\Sigma}_{SKC}$ is that this adjusted sandwich estimator was obtained under the assumption that the form of the covariance matrix was *correctly specified*; however, Kauermann and Carroll (2001) demonstrated that their estimator is "rather robust against misspecified covariances."

Lu et al. (2007) made a thorough comparison of the finite sample properties of the sandwich estimators via simulations for equally sized clusters and categorical covariates. They showed that it could be appropriate to fit the usual sandwich estimator $\widehat{\Sigma}_S$ if the number of clusters or subjects is greater than 40. For moderate to large cluster sizes, they suggested the application of $\widehat{\Sigma}_{SMR}$; for studies with small clusters sizes (≤ 10), they suggested that $\widehat{\Sigma}_{SKC}$ might be preferable for studies with cluster level covariates, while $\widehat{\Sigma}_{SMR}$ might be best for within-cluster covariates. However, Lu et al. (2007) also noted that more research is needed for unequal cluster sizes, continuous covariates, and for high leverage values that are known to cause difficulty with respect to undercoverage of confidence intervals.

It is also of interest to note that the GEE model-based variance estimator $\widehat{\Sigma}_M$, like the sandwich-based estimator $\widehat{\Sigma}_S$, can also underestimate the true variance when the sample sizes are small. However, unlike the sandwich-based estimators, whose downward bias tends to be due to residuals that are too small, the downward bias for the model-based estimator is related to underestimation of α. Teerenstra et al. (2010) overcame the negative bias in the model-based estimator in part by using improved estimators of α that they obtained using the "matrix adjusted" estimating equation (MAEE) approach described by Preisser et al. (2003b).

Finally, we note that if we specify a canonical link function, then we can achieve further simplification in all the estimated covariance matrices discussed in this section. For canonical link functions, $\frac{\partial \mu_{ij}}{\partial \beta_k} = 1/\phi Var(y_{ij})x_{ijk}$, so that $D_i = \frac{\partial \mu_i}{\partial \beta} = A_i X_i$; Substitution into Equation (2.25) therefore yields the following simplified version of the model based estimator for a canonical link function:

$$\widehat{Cov}_M(\widehat{\beta}) = \widehat{\phi}_{GEE} \widehat{S}_m^{-1}, \tag{2.34}$$

where

$$\widehat{S}_m = \sum_{i=1}^{m} X_i' A_i^{1/2} R_i^{-1}(\widehat{\alpha}) A_i^{1/2} X_i.$$

In addition, the sandwich estimator equation (2.24) covariance matrix has the following form for the canonical link function:

$$\widehat{Cov}_R(\widehat{\beta}) = \widehat{S}_m^{-1} \left\{ \sum_{i=1}^{m} X_i' A_i^{1/2} R_i^{-1}(\widehat{\alpha}) Z_i(\widehat{\beta}) Z_i'(\widehat{\beta}) R_i^{-1}(\widehat{\alpha}) A_i^{1/2} X_i \right\} \widehat{S}_m^{-1}. \tag{2.35}$$

Similar simplifications would also be possible for the other estimators provided in this section.

2.4   Application for Obesity Study Provided in Chapter 1

Here we demonstrate implementation of GEE for analysis of the longitudinal study in obesity described in Section 1.5.1 using the **xtgee** command in Stata 12.0. Commands for loading the obesity data and replicating all the examples in this section in R, MATLAB, and SAS are provided on the book's website: https://dbe.med.upenn.edu/biostat-research/Book-QLS. The software for implementing GEE in MATLAB is described in Ratcliffe and Shults (2008).

We fit a regression model with an identity link function to regress change in BMI z-score since baseline on *month* and *month*-squared, while adjusting for the correlation between repeated measurements on a subject via application of the correlation structures described in Section 2.3.3.

We first open up the dataset, generate, and then tabulate a lag variable that represents the months between consecutive measurements on a subject during the first 36 months following transplant:

```
. sort id month
. by id: gen lag = month - month[_n-1] if _n>1
. tab lag if month <= 36
```

lag	Freq.	Percent	Cum.
2	81	22.13	22.13
3	77	21.04	43.17
5	6	1.64	44.81
6	76	20.77	65.57
9	11	3.01	68.58
11	2	0.55	69.13
12	105	28.69	97.81
18	4	1.09	98.91
24	4	1.09	100.00
Total	366	100.00	

As can be seen from the above tabulation, measurements were unequally spaced during the first 3 years following renal transplant. As noted in Section 1.5.1, each patient had measurement times that were a subset of the common set of measurement times $\{1, 3, 6, 12, 24, 36\}$. Next, we generate the change variable that represents change in BMI z-score since baseline:

```
. use random_small
. gen change = bmiz - basebmiz
. gen month2 = month^2
```

Application of AR(1) working correlation structure: We start by fitting GEE with a model-based covariance matrix. (Note the use of the option *force* that forces Stata to fit the AR(1) structure even though the measurements are not equally spaced in time.)

```
. xtgee change month month2  if month<=36, i(id) t(month) force c(AR 1)

Iteration 1: tolerance = .01810285
Iteration 2: tolerance = .00043698
Iteration 3: tolerance = 8.044e-06
Iteration 4: tolerance = 1.483e-07

GEE population-averaged model                    Number of obs     =    466
Group and time vars:              id month        Number of groups  =    100
```

```
Link:                            identity    Obs per group: min =              2
Family:                          Gaussian                   avg =            4.7
Correlation:                        AR(1)                   max =              6
                                             Wald chi2(2)        =           2.59
Scale parameter:                  .8242595   Prob > chi2         =         0.2739

-------------------------------------------------------------------------------
    change |      Coef.   Std. Err.       z    P>|z|     [95% Conf. Interval]
-----------+-------------------------------------------------------------------
     month |   .0018156   .0077806     0.23    0.815    -.0134341     .0170652
    month2 |  -.0001528   .0001925    -0.79    0.427    -.0005301     .0002245
     _cons |   .8364866   .0919279     9.10    0.000     .6563112     1.016662
-------------------------------------------------------------------------------
```

As can be seen from the above output, the total number of measurements (Number of obs) was 466, while the number of subjects (Number of groups) was 100; in addition, there were between 2 and 6 measurements per subject, with an average 4.7 measurements per subject. Note also that the value of the criterion for convergence (tolerance) decreases in value with each iteration and is very small at Iteration 4, which was the final iteration prior to convergence in the GEE estimation procedure. (In order to save space, in the remaining examples we will only list the tolerance at the final iteration.)

Next, we display the estimated correlation structure. Note that $\hat{\alpha}_{GEE} = 0.8634$ below, which indicates that the measurements within subjects are highly correlated.

. xtcorr

Estimated within-id correlation matrix R:

```
         c1       c2       c3       c4       c5       c6
r1   1.0000
r2   0.8634   1.0000
r3   0.7455   0.8634   1.0000
r4   0.6437   0.7455   0.8634   1.0000
r5   0.5557   0.6437   0.7455   0.8634   1.0000
r6   0.4798   0.5557   0.6437   0.7455   0.8634   1.0000
```

Note that the dimension of estimated correlation matrix is 6, which is the largest number of measurements observed on a subject. Next, we fit GEE with the robust sandwich covariance matrix: (In the remaining examples for this section, we only display part of the output.)

. xtgee change month month2 if month<=36, i(id) t(month) force c(AR 1) robust

Iteration 4: tolerance = 1.483e-07
```
                                    (Std. Err. adjusted for clustering on id)
-------------------------------------------------------------------------------
           |             Semirobust
    change |      Coef.   Std. Err.       z    P>|z|     [95% Conf. Interval]
-----------+-------------------------------------------------------------------
     month |   .0018156   .0097793     0.19    0.853    -.0173515     .0209826
```

```
month2 |  -.0001528   .0002334   -0.65   0.513   -.0006102    .0003046
 _cons |   .8364866   .0725075   11.54   0.000    .6943746    .9785986
----------------------------------------------------------------------------
```

Note that the regression coefficients for *month* and *month*-squared did not differ significantly from zero, for both the model-based and robust sandwich covariance matrix. We will revisit this example in Chapter 5 when we implement QLS with a working correlation structure that, in contrast to the AR(1) structure applied here, is appropriate for irregularly timed measurements.

If we fit the same model in MATLAB, SAS, and R, the estimates for α would differ slightly between the different packages between Stata, R, and SAS, while the results are identical for MATLAB and SAS; as mentioned earlier, SAS and MATLAB implement the same estimate for α for the AR(1) structure, while the form of the estimator differs slightly between the other packages for GEE.

For comparison, we will also fit the other structures that were described in Section 2.3.3.

Application of exchangeable working correlation structure: First, we fit the model-based covariance matrix:

```
Iteration 2: tolerance = 5.476e-07


GEE population-averaged model          Number of obs      =        466
Group variable:                    id  Number of groups   =        100
Link:                        identity  Obs per group: min =          2
Family:                      Gaussian                 avg =        4.7
Correlation:             exchangeable                 max =          6
                                       Wald chi2(2)       =      18.46
Scale parameter:              .8112636  Prob > chi2       =     0.0001


----------------------------------------------------------------------------
   change |     Coef.   Std. Err.      z    P>|z|    [95% Conf. Interval]
----------+-----------------------------------------------------------------
    month |   .016759   .0069773    2.40   0.016    .0030838    .0304341
   month2 | -.0006465   .0001956   -3.31   0.001   -.0010298   -.0002632
    _cons |  .8714932   .0880598    9.90   0.000    .6988991    1.044087
----------------------------------------------------------------------------


. xtcorr

Estimated within-id correlation matrix R:

         c1      c2      c3      c4      c5      c6
r1   1.0000
r2   0.7605  1.0000
r3   0.7605  0.7605  1.0000
r4   0.7605  0.7605  0.7605  1.0000
r5   0.7605  0.7605  0.7605  0.7605  1.0000
r6   0.7605  0.7605  0.7605  0.7605  0.7605  1.0000
```

Next, we fit the sandwich covariance matrix.

```
. xtgee change month month2 if month<=36, i(id) t(month)  c(exc) robust

Iteration 2: tolerance = 5.476e-07
```

```
                                     (Std. Err. adjusted for clustering on id)
-----------------------------------------------------------------------------
            |             Semirobust
   change   |    Coef.    Std. Err.      z     P>|z|     [95% Conf. Interval]
------------+----------------------------------------------------------------
    month   |  .016759    .0090101     1.86   0.063    -.0009006     .0344185
   month2   | -.0006465   .000221     -2.93   0.003    -.0010796    -.0002133
    _cons   |  .8714932   .0733427    11.88   0.000     .7277441    1.015242
-----------------------------------------------------------------------------
```

The GEE procedure converged within two iterations for this structure. Also, the estimated correlation is again large, with $\widehat{\alpha}_{GEE} = 0.7605$. In addition, the regression coefficient for *month* differed significantly from zero only for the model-based covariance matrix.

Application of tri-diagonal working correlation structure:

```
. xtgee change month month2 if month<=36, i(id) t(month) force  c(sta 1) robust

Iteration 4: tolerance = 2.950e-07
                                     (Std. Err. adjusted for clustering on id)
-----------------------------------------------------------------------------
            |             Semirobust
   change   |    Coef.    Std. Err.      z     P>|z|     [95% Conf. Interval]
------------+----------------------------------------------------------------
    month   |  .0041207   .0108347     0.38   0.704    -.0171149     .0253564
   month2   | -.0001179   .0002861    -0.41   0.680    -.0006787     .0004429
    _cons   |  .8693962   .0717645    12.11   0.000     .7287403    1.010052
-----------------------------------------------------------------------------
convergence not achieved
r(430);
```

Note that we obtained an error message that convergence was not achieved, although the final iteration was 4 with a small value of the criterion for convergence at this final iteration. This suggests that the estimated regression parameter values in iteration 4 were close to the estimated values in iteration 3. Next we print the estimated correlation matrix to determine what may have been the cause of the error message:

```
. xtcorr


Estimated within-id correlation matrix R:

          c1       c2       c3       c4       c5       c6
r1    1.0000
r2    0.8583   1.0000
r3    0.0000   0.8583   1.0000
r4    0.0000   0.0000   0.8583   1.0000
r5    0.0000   0.0000   0.0000   0.8583   1.0000
r6    0.0000   0.0000   0.0000   0.0000   0.8583   1.0000
```

As described in Section 2.3.3, the feasible interval for α for the tri-diagonal structure is $(-1/c_{max}, 1/c_{max})$, where $c_{max} = 2 \sin \left(\frac{\pi[n_{max}-1]}{2[n_{max}+1]} \right)$ and n_{max} is the maximum value of n_i over $i = 1, 2, \ldots, m$. We noted earlier that there were between 2 and 6 measurements on each subject, so that $n_{max} = 6$ for this example. We can use Stata to find the feasible interval for α as follows:

```
. global n_m = 6
. display 1/(2*sin( _pi*($n_m - 1)/(2*($n_m + 1)))))
.55495813
```

The feasible interval for α for this example is therefore approximately $(-0.55, 0.55)$. Our estimated value of α is 0.8583, which is not contained in $(-0.55, 0.55)$, so that our estimated correlation matrix is not positive definite. We can also confirm that the estimated correlation matrix is not positive definite by obtaining the eigenvalues of the estimated correlation matrix and checking that they are not all positive:

```
. matrix R = e(R)
. matrix symeigen eigenvectors eigenvalues = R
. matrix list eigenvalues

eigenvalues[1,6]
            e1         e2         e3         e4          e5          e6
r1    2.5466329  2.0703032  1.3819868  .61801322  -.07030318  -.54663291
```

As can be seen from the above listing of eigenvalues, two of the eigenvalues are negative, so that the estimated correlation matrix is not positive definite.

Application of unstructured working correlation matrix:

```
. xtgee change month month2 if month<=36, i(id)  t(month)   c(uns)  robust

Iteration 100: tolerance = .00621817

GEE population-averaged model          Number of obs      =      466
Group and time vars:          id month  Number of groups   =      100
Link:                         identity  Obs per group: min =        2
Family:                       Gaussian                 avg =      4.7
Correlation:              unstructured                 max =        6
                                        Wald chi2(2)       =     2.12
Scale parameter:                .8228771  Prob > chi2      =   0.3472

                                (Std. Err. adjusted for clustering on id)
------------------------------------------------------------------------
             |             Semirobust
      change |    Coef.    Std. Err.     z    P>|z|    [95% Conf. Interval]
-------------+----------------------------------------------------------
       month |  .0328574   .0229045    1.43   0.151   -.0120346   .0777494
      month2 | -.0008194   .0006052   -1.35   0.176   -.0020055   .0003668
       _cons |   .78502    .1149069    6.83   0.000    .5598065   1.010233
------------------------------------------------------------------------
convergence not achieved
r(430);
```

As described earlier, the unstructured matrix is typically applied when we have a common set of measurement times and a relatively small number of measurements per subject. For this study, the measurement times are a subset of a common set of measurement times but they vary between subjects. Of the 100 patients in this study, the percentage of subjects at months 1, 3, 6, 12, 24, and 36 was 89 percent, 91 percent, 84 percent, 89 percent, 68 percent, and 45 percent, respectively. There was therefore a high percentage of missing data at months 24 and 36. In addition, the number of correlation parameters involved in the unstructured matrix was relatively large (15). The failure to converge for the unstructured matrix was therefore not surprising for this example.

It is also interesting that, unlike the previous implementation of the tri-diagonal structure, the criterion for convergence was relatively large at the 100th iteration. We next obtain the estimated correlation matrix that was obtained at the final iteration and check to see if it was positive definite:

```
. xtcorr

Estimated within-id correlation matrix R:

         c1       c2       c3       c4       c5       c6
r1   1.0000
r2   0.5606   1.0000
r3   0.4624   0.7311   1.0000
r4   0.5806   0.6609   0.7044   1.0000
r5   0.8628   0.7381   0.7571   1.0000   1.0000
r6   0.6720   0.6276   0.4820   0.9692   1.0000   1.0000

. matrix R = e(R)
. matrix symeigen eigenvectors eigenvalues = R
. matrix list eigenvalues

eigenvalues[1,6]
          e1         e2         e3         e4         e5         e6
r1   4.649412  .69424671  .48504456  .28475948  -.02132088  -.09214183
```

As for the tri-diagonal structure, the unstructured correlation matrix that was obtained at the final iteration is not positive definite because two of its estimated eigenvalues are negative.

Application of identity working correlation structure: Application of an identity structure, coupled with a sandwich covariance matrix, is sometimes suggested as an approach to adjust for correlation in the data; although the identity structure assumes that $\alpha = 0$, implementation of the sandwich matrix provides a correction if the assumption of zero correlations is wrong.

```
. xtgee change month month2 if month<=36, i(id) t(month)  c(Ind) robust

Iteration 1: tolerance = 2.677e-15
                                   (Std. Err. adjusted for clustering on id)
-----------------------------------------------------------------------------
             |               Semirobust
      change |    Coef.    Std. Err.      z     P>|z|     [95% Conf. Interval]
-------------+---------------------------------------------------------------
       month |    .02074    .0097208     2.13   0.033     .0016876    .0397925
      month2 |  -.0007411    .0002557    -2.90   0.004    -.0012422   -.0002401
       _cons |   .8663075    .0741369    11.69   0.000     .7210017    1.011613
-----------------------------------------------------------------------------
```

Note that we achieved convergence in one iteration because we do not need multiple iterations that cycle between updating the estimates of α and β when we are fitting an identity structure that assumes $\alpha = 0$. Next we obtain the estimated correlation matrix to confirm that it is indeed an identity structure.

```
. xtcorr
```

```
Estimated within-id correlation matrix R:

          c1        c2        c3        c4        c5        c6
r1    1.0000
r2    0.0000    1.0000
r3    0.0000    0.0000    1.0000
r4    0.0000    0.0000    0.0000    1.0000
r5    0.0000    0.0000    0.0000    0.0000    1.0000
r6    0.0000    0.0000    0.0000    0.0000    0.0000    1.0000
```

Comments on implementation of the different structures: Error messages indicated that convergence was not achieved for the tri-diagonal and unstructured matrices, although the final criterion for convergence (at iteration 4) was small for the tri-diagonal structure, albeit with a final estimated correlation matrix that was not positive definite.

The results differed according to the choice of working correlation structure. The estimates of the regression coefficients for *month*, *month*-squared, and the constant were approximately 0.02, −0.001, and 0.87, respectively, for both the exchangeable and identity structures; however, all three regression coefficients differed significantly from zero at a 0.05 criterion for significance only for the identity structure and for the exchangeable structure with the model-based estimate of the covariance. (For the exchangeable structure with sandwich covariance, the p-values for *month*, *month*-squared, and constant were 0.06, 0.003, and <0.0005, respectively.) The estimates for *month* and *month*-squared were much smaller for the AR(1) structure (0.002 and < -0.0005, respectively); in addition, neither *month* nor *month*-squared differed significantly from zero for this structure. These varying results provide motivation for assessing the fit of the models, as we will do in Chapter 8. As mentioned above, we will also implement a structure that is appropriate for unequally spaced measurements, in Chapter 5.

2.5   Exercises

Exercise 2.1 *This Exercise is designed to review some important concepts of linear models that were briefly described in Section 2.2.1. We assume the regression model is full-rank, so that the design matrix has full column rank.*

   a. *Show why minimization of the error sum of squares and application of the principle of maximum likelihood leads to minimization of the same objective function, when the outcome variable is assumed to be normally distributed.*

   b. *Next, minimize the objective function with respect to the regression parameter to obtain the estimating equation (normal equations) for β. Prove that the solution to the normal equations does indeed minimize the error sum of squares.*

   c. *What is the distribution of the solution $\widehat{\boldsymbol{\beta}}$ to the normal equations?*

Exercise 2.2 *Show that the normal and Poisson distributions can be expressed in the form of the general exponential family in Equation (2.1), for the characteristics given in Table 2.1. Next, obtain the mean and variance of each distribution using $\mu_{ij} = b'(\theta_{ij})$ and $Var(y_{ij}) = b''(\theta_{ij})\phi$, respectively.*

Exercise 2.3 *Apply the principle of maximum likelihood for estimation of $\boldsymbol{\beta}$ in the exponential distribution (2.1) when $\theta_i = \boldsymbol{x}_i'\boldsymbol{\beta}$.*

   a. *Obtain the log-likelihood that was provided in Equation (2.3).*

   b. *Next, obtain the maximum likelihood estimating equation for $\boldsymbol{\beta}$ that was provided in Equation (2.4).*

Exercise 2.4 *Show that Equation (2.5) and (2.6) are equivalent.*

Exercise 2.5 *Re-express Equation (2.6) to obtain the equivalent GLM estimating equation (2.7) for $\boldsymbol{\beta}$.*

Exercise 2.6 *Suppose that one measurement is collected on each of m subjects. Show that the GEE estimating equation (2.9) then reduces to the GLM estimating equation (2.6).*

Exercise 2.7 *Suppose that $\boldsymbol{\beta}$ is known in the moment estimates for GEE, so that $\widehat{\boldsymbol{\beta}} = \boldsymbol{\beta}$.*

   a. *Find the expected value of Equation (2.18).*

   b. *Next, find the expected value of Equation (2.17) for the equicorrelated structure.*

   c. *Next, find the expected value of Equation (2.19) for the AR(1) and tri-diagonal structures.*

   d. *Next, find the expected value of Equation (2.20) for the unstructured matrices.*

Exercise 2.8 *For canonical link functions, $\frac{\partial \mu_{ij}}{\partial \beta_k} = 1/\phi Var(y_{ij})x_{ijk}$, so that $D_i = \frac{\partial \mu_i}{\partial \beta} = \frac{1}{\phi}A_iX_i$.*

   a. *Show that substitution into Equation (2.25) therefore yields the simplified version of the model based estimator for a canonical link function that is provided in Equation (2.34).*

$$\widehat{Cov}_M(\widehat{\beta}) = \widehat{\phi}_{GEE}\widehat{S}_m^{-1},$$

   where

$$\widehat{S}_m = \sum_{i=1}^{m} X_i'A_i^{1/2}R_i^{-1}(\widehat{\alpha})A_i^{1/2}X_i.$$

   b. *In addition, show that the sandwich (2.24) covariance matrix has the form for the canonical link function that is provided in (2.35).*

Part II

Quasi-Least Squares Theory and Applications

Chapter 3

History and Theory of Quasi-Least Squares Regression

Quasi-least squares (QLS) is a two-stage computational approach for estimation of the correlation parameters that is in the framework of generalized estimating equations (GEE). QLS was developed in a series of papers. Stage one was proposed for balanced data ($n_i = n \; \forall \; i$) in Chaganty (1997) and was then extended for unbalanced and unequally spaced data in Shults (1996) and Shults and Chaganty (1998). A second stage of the QLS procedure was then provided in Chaganty and Shults (1999). In this chapter, we provide a more detailed description of the development of QLS than was possible in previous manuscripts, which typically had little room for more than a brief summary of the approach and of the manuscripts on which it was based. To streamline our descriptions, we describe how QLS evolved for one particular correlation structure, the first-order autoregressive or AR(1) structure. Other correlation structures will be presented in subsequent chapters. Our focus on the AR(1) structure here is solely for clarity of exposition as we describe the results contained in the original manuscripts on QLS.

Throughout this chapter we use the more general notation that is appropriate for correlated data, as we did for our discussion of GEE in the previous chapter. In Section 2.3.3 we defined the feasible interval for a correlation matrix to be the interval on which α will yield a positive definite correlation matrix. In this chapter we will refer to estimators of α as feasible (or positive definite) if they always yield a positive definite estimate of the working correlation structure. We will refer to an estimate (the value taken by an estimator for a particular dataset) as feasible, or positive definite, if it takes value in the feasible (positive definite) interval for a particular structure.

It will be important to distinguish between feasible (positive definite) versus consistent estimators. For example, we shall see in this chapter that the QLS stage one and two estimators for the AR(1) structure are feasible, so that the resultant estimated correlation matrices will always be positive definite. This is in contrast to an estimator that is consistent but not necessarily feasible, which will guarantee a positive definite matrix asymptotically but not necessarily for small (or even large) sample sizes.

3.1   Why QLS is a "Quasi"-Least Squares Approach

QLS is referred to as a "quasi"-least-squares (LS) approach because its estimates do not fully conform to the principle of least squares (LS). We explore the relationship between the QLS and a LS approach below, following discussions similar to those provided by Dunlop (1994), who explored the link between LS, GLM, and GEE. Our strategy will be to generalize the objective function for LS estimation of $\boldsymbol{\beta}$ for a classical linear regression model to the more general framework for QLS and GEE; consideration of the generalized objective function for LS estimation of $\boldsymbol{\beta}$ will then yield "quasi"-least squares estimates of the parameters:

First, consider the classical linear regression model (see Section 2.2.1 in the previous chapter) that assumes the correlation between any two measurements y_{ij} and y_{ik} $(j \neq k)$ is zero, for $i = 1, 2, \ldots, m$ and $j, k \in \{1, 2, \ldots, n_i\}$:

$$Y = X\boldsymbol{\beta} + e; \; E(e) = 0; \text{ and } Cov(e) = \phi I, \tag{3.1}$$

where $Y = (y_{11}, \ldots, y_{1n_1}, \ldots, y_{m1}, \ldots, y_{mn_m})'$; $e = (e_{11}, \ldots, e_{1n_1}, \ldots, e_{m1}, \ldots, e_{mn_m})'$; $\boldsymbol{\beta} = (\beta_1, \ldots, \beta_p)'$ and X is the $N \times p$ covariance matrix, for which $(x_{ij1}, x_{ij2}, \ldots, x_{ijp})$ is the $\sum_{l=1}^{i-1} n_l + j^{th}$ row (for $n_o = 0$), and $N = \sum_{i=1}^{m} n_i$.

The objective function for the LS approach to estimation of $\boldsymbol{\beta}$ is the sum of squared residuals,

$$Q(\beta) \;=\; \sum_{i=1}^{m} \sum_{j=1}^{n_i} e_{ij}^{2} \tag{3.2}$$

$$=\; (Y - X\boldsymbol{\beta})' (Y - X\boldsymbol{\beta}) \tag{3.3}$$

$$=\; (Y - \boldsymbol{\mu})' (Y - \boldsymbol{\mu}), \tag{3.4}$$

where $\mu = E(Y)$.

The LS approach for estimation of $\boldsymbol{\beta}$ requires the minimization of Q with respect to $\boldsymbol{\beta}$, so that the estimating equations for $\boldsymbol{\beta}$ (the *normal equations* $X'X\boldsymbol{\beta} = X'Y$) are obtained by simplifying the following equation: (See Exercise 3.1)

$$\frac{\partial Q}{\partial \boldsymbol{\beta}} = 0. \tag{3.5}$$

Next, consider the situation in which the data are correlated, but with covariances that are fixed and known (up to parameter ϕ). Our assumed model is now

$$Y = X\boldsymbol{\beta} + e; \; E(e) = 0; \text{ and } Cov(e) = \phi V, \tag{3.6}$$

where V is a known and positive-definite $N \times N$ matrix. Although the errors in Equation (3.6) are correlated, it is well known that we can transform the data so that the regression model (3.6) with positive-definite covariance matrix ϕV for V fixed and known can be expressed as an equivalent classical regression model (3.1) with covariance matrix ϕI: Pre-multiplying the left- and right-hand sides of Equation (3.6) yields the following transformed model:

$$V^{-1/2} Y = V^{-1/2} X\boldsymbol{\beta} + V^{-1/2} e. \tag{3.7}$$

Model (3.7) can then be expressed as

$$Y^* = X^*\boldsymbol{\beta} + e^*, \tag{3.8}$$

where $Y^* = V^{-1/2}Y$; $X^* = V^{-1/2}X$; and $e^* = V^{-1/2}e$.

Model (3.8) is equivalent to the classical regression model (3.1) because $E(e^*) = V^{-1/2}\mathbf{0} = \mathbf{0}$ and the covariance matrix of the transformed errors is diagonal: $Cov(e^*) = Cov(V^{-1/2}e) = \phi V^{-1/2}VV^{-1/2} = \phi I$. The practical significance of obtaining the transformed model (3.8) lies in the fact that results obtained for the classical linear model (3.1) can be directly applied for this model; in other words, there is no need to develop new theory for correlated measurements when V is fixed and known. In particular, the objective function for LS estimation of $\boldsymbol{\beta}$ can be expressed as follows:

$$
\begin{aligned}
Q^*(\beta) &= (Y^* - \boldsymbol{\mu}^*)'(Y^* - \boldsymbol{\mu}^*) & (3.9) \\
&= (Y^* - X^*\boldsymbol{\beta})'(Y^* - X^*\boldsymbol{\beta}) & (3.10) \\
&= (Y - X\boldsymbol{\beta})'V^{-1}(Y - X\boldsymbol{\beta}) & (3.11) \\
&= (Y - \boldsymbol{\mu})'V^{-1}(Y - \boldsymbol{\mu}). & (3.12)
\end{aligned}
$$

Minimization of $Q^*(\beta)$ with respect to $\boldsymbol{\beta}$ yields the weighted least squares estimating equations $X'V^{-1}X\beta = X'V^{-1}Y$ ((Miller et al., 1993; Stokes et al., 2000)). As described in Chaganty (1997), QLS extends consideration of the objective function $Q^*(\beta)$ in Equation (3.9) for weighted least squares when the covariance matrix is known to $Q^*(\alpha, \beta)$ for least squares estimation of $\boldsymbol{\beta}$ and unknown correlation parameters α in the more general framework of GEE. However, note that in this more general setting (see Section 2.3 of the previous chapter and Section 2.3.4 in particular), not only are the y_{ij} correlated (with correlations that may depend on unknown correlation parameters α), but the covariance matrix ϕV may depend on the unknown regression parameter $\boldsymbol{\beta}$, in addition to the parameter α. For example, if we consider a Poisson model (with log link function), then $Cov(Y) = \phi V = \phi diag(V_1, V_2, \ldots, V_m)$, where $V_i = V_i(\alpha, \boldsymbol{\beta}) = A_i^{1/2}(\boldsymbol{\beta})R_i(\alpha)A_i^{1/2}(\boldsymbol{\beta})$ and $A_i^{1/2}(\boldsymbol{\beta}) = diag(\sqrt{\mu_{i1}}, \sqrt{\mu_{i2}}, \ldots, \sqrt{\mu_{in_i}})$ for $\mu_{ij} = \exp(x_{ij}'\boldsymbol{\beta})$. The covariance matrix therefore depends on $\boldsymbol{\beta}$ for the Poisson model, as it does for any model for which the variances $Var(y_{ij})$ depend on $\boldsymbol{\beta}$; for these models, Q^* can be expressed as

$$
\begin{aligned}
Q^*(\alpha, \beta) &= (Y - \boldsymbol{\mu})'V^{-1}(\alpha, \boldsymbol{\beta})(Y - \boldsymbol{\mu}) & (3.13) \\
&= \sum_{i=1}^{m}(Y_i - \boldsymbol{\mu}_i)'V_i^{-1}(\alpha, \boldsymbol{\beta})(Y_i - \boldsymbol{\mu}_i) & (3.14) \\
&= \sum_{i=1}^{m}(Y_i - \boldsymbol{\mu}_i)'A_i^{-1/2}(\boldsymbol{\beta})R_i^{-1}(\alpha)A_i^{-1/2}(\boldsymbol{\beta})(Y_i - \boldsymbol{\mu}_i), & (3.15)
\end{aligned}
$$

where V_i^{-1} is expressed as $V_i^{-1}(\alpha, \boldsymbol{\beta})$ to reflect the fact that the covariances may depend on both the regression and correlation parameters. Shults and Chaganty (1998) referred to $Q^*(\alpha, \beta)$ as the "generalized error sum of squares."

It could be argued that $Q^*(\alpha, \boldsymbol{\beta})$ has appeal as an objective function for LS estimation of $\boldsymbol{\beta}$ in the more general setup for GEE because it is a direct extension of the objective function for LS estimation of the parameters in the linear model (3.8) for correlated data, when the correlations are known. The QLS approach for LS estimation of $\boldsymbol{\beta}$ that was described in Chaganty (1997) views Equation (3.13) as a function of α, $\boldsymbol{\beta}$, and $\boldsymbol{\beta}^*$, expressed as $Q^*(\alpha, \boldsymbol{\beta}, \boldsymbol{\beta}^*)$, where the expected value for each $\boldsymbol{Y}_i$, $E(\boldsymbol{Y}_i) = \boldsymbol{\mu}_i(\boldsymbol{\beta})$, is a function of $\boldsymbol{\beta}$ and the covariance matrix of each $\boldsymbol{Y}_i$, $Cov(\boldsymbol{Y}_i) = A_i^{1/2}(\boldsymbol{\beta}^*)R_i(\alpha)A_i^{1/2}(\boldsymbol{\beta}^*)$, is a function of $\boldsymbol{\beta}^*$. Next, QLS differentiates $Q^*(\alpha, \boldsymbol{\beta}, \boldsymbol{\beta}^*)$ with respect to $\boldsymbol{\beta}$; equates the resulting estimating function with zero; and then sets $\boldsymbol{\beta}^* = \boldsymbol{\beta}$. In other words, QLS treats $\boldsymbol{\beta}$ as fixed and known in the covariance matrix of $\boldsymbol{Y}_i$, for $i = 1, 2, \ldots, m$.

The QLS approach with respect to estimation of $\boldsymbol{\beta}$ can also be described in the following equivalent, and perhaps simpler, fashion. First, differentiate $Q^*(\alpha, \boldsymbol{\beta})$ with respect to $\boldsymbol{\beta}$ and simplify the LS estimating equation $\frac{\partial}{\partial \boldsymbol{\beta}} Q^*(\alpha, \boldsymbol{\beta}) = \boldsymbol{0}$ to obtain an estimating equation with two terms:

$$\left(\frac{\partial \boldsymbol{\mu}}{\partial \boldsymbol{\beta}}\right)' \frac{\partial Q^*(\alpha, \boldsymbol{\beta})}{\partial \boldsymbol{\mu}} + \sum_{i=1}^{m} (\boldsymbol{Y}_i - \boldsymbol{\mu}_i)' \frac{\partial V_i^{-1}(\alpha, \boldsymbol{\beta})}{\partial \boldsymbol{\beta}} (\boldsymbol{Y}_i - \boldsymbol{\mu}_i) = \boldsymbol{0}. \qquad (3.16)$$

Next, discard the second term of the estimating equation (3.16) and then simplify. It is straightforward to show (see Problem 3.2) that the first term of the estimating function in Equation (3.16) is the GEE estimating function for $\boldsymbol{\beta}$. The second term will be non-zero, unless we consider the identity link function, in which case the covariance matrix V is not a function of $\boldsymbol{\beta}$, so that each $\frac{\partial V_i^{-1}(\alpha, \boldsymbol{\beta})}{\partial \boldsymbol{\beta}} = \frac{\partial V_i^{-1}(\alpha)}{\partial \boldsymbol{\beta}}$ is a matrix of zeros. The QLS estimation procedure for $\boldsymbol{\beta}$ that was described in Chaganty (1997) and summarized here is therefore equivalent to differentiating the generalized error sum of squares with respect to $\boldsymbol{\beta}$ and then discarding the second term, so that the final estimating equation for $\boldsymbol{\beta}$ is identical to the estimating equation for GEE.

Note: Dunlop (1994) had observed that even for generalized linear models, minimization of $Q^*(\beta)$ with respect to $\boldsymbol{\beta}$ would yield an estimating function that is more complex than the estimating function for GLM. For example, differentiation of the generalized error sum of squares for a logistic or Poisson model for independent measurements will also yield two non-zero terms (represented by the left-hand side of Equation (3.16) for $\alpha = 0$), the first of which is the GLM estimating function for $\boldsymbol{\beta}$.

In contrast to its "partial minimization approach" with respect to $\boldsymbol{\beta}$, *stage one* of QLS fully differentiates the generalized error sum of squares with respect to α, as described in detail in subsequent sections. We will see that QLS requires two stages of estimation of α. Because QLS does not fully apply the principle of least squares when minimizing the objective function $Q^*(\alpha, \boldsymbol{\beta})$ with respect to $\boldsymbol{\beta}$, the stage one estimates do not fully conform to the principle of least squares. As noted on page 42 of Chaganty (1997), "Since the estimates do not fully conform to the principle of (generalized) least squares, it is reasonable to call our estimates $\widehat{\beta}$ and $\widehat{\alpha}$, 'quasi-least squares estimates' of $\boldsymbol{\beta}$ and α, respectively." (As shown above, if we fully applied the principle of least squares, the estimating function for $\boldsymbol{\beta}$ would be a function of two terms that differs from the LS estimating function for the non-identity link function.)

Note on semantics: Because QLS and GEE both implement the GEE estimating equation for $\boldsymbol{\beta}$, more precise names for the two approaches might be *quasi and least squares* versus *quasi and moment estimation*, respectively. For QLS, *quasi* and *least squares* reflect that a "quasi"-LS approach is used for estimation of $\boldsymbol{\beta}$, while a full LS approach is used for estimation of α (in stage one of the procedure, as will be discussed in the next section). For GEE, *quasi* and *moment* estimation reflect that a "quasi"-LS approach is used for estimation of $\boldsymbol{\beta}$, while moment estimation is typically applied for estimation of α.

3.2   The Least Squares Approach Employed in Stage One of QLS for Estimation of α

In the previous section we saw that QLS is referred to as a "quasi"-least-squares approach because (like GEE) it does not employ a full least squares approach with respect to estimation of $\boldsymbol{\beta}$. However, QLS does employ a least squares approach with respect to estimation of the correlation parameter α in the first stage of its estimation procedure for α.

Stage one of QLS extends Dunlop's consideration of LS estimation of $\boldsymbol{\beta}$ for GLM and GEE to the correlation parameters. While Dunlop (1994) considered minimizing the objective function $Q^*(\boldsymbol{\beta})$ in Equation (3.9) with respect to $\boldsymbol{\beta}$ for GLM (see the second paragraph on page 301 of Dunlop (1994)), stage one of QLS minimizes the generalized error sum of squares $Q^*(\alpha, \boldsymbol{\beta})$ in Equation (3.13) with respect to $\boldsymbol{\alpha}$. As described in Chaganty (1997) for balanced data ($n_i = n \; \forall \; i$), stage one of QLS therefore obtains an estimate $\widehat{\alpha}_{QONE}$ of the correlation parameter α by obtaining the solution for α to

$$\frac{\partial}{\partial \alpha} Q^*(\alpha, \boldsymbol{\beta}) = \mathbf{0},$$

which can be expressed as

$$\sum_{i=1}^{m} Z_i'(\boldsymbol{\beta}) \frac{\partial R_i^{-1}(\alpha)}{\partial \alpha} Z_i(\boldsymbol{\beta}) = \mathbf{0}, \tag{3.17}$$

where

$$Z_i(\boldsymbol{\beta}) = A_i^{-1/2}(\boldsymbol{\beta})(\boldsymbol{Y}_i - \boldsymbol{\mu}_i) \tag{3.18}$$

is the vector of Pearson residuals on subject i. The QLS stage one estimate $\widehat{\alpha}_{QONE}$ therefore satisfies the following equation:

$$\sum_{i=1}^{m} Z_i'(\boldsymbol{\beta}) \frac{\partial R_i^{-1}(\alpha)}{\partial \alpha} \bigg|_{\alpha = \widehat{\alpha}_{QONE}} Z_i(\boldsymbol{\beta}) = \mathbf{0}. \tag{3.19}$$

When we consider the solution of Equation (3.17) as m approaches ∞, it will also be helpful to express Equation (3.17) in the following equivalent form:

$$\frac{1}{m}\sum_{i=1}^{m}Z_i'(\boldsymbol{\beta})\frac{\partial R_i^{-1}(\alpha)}{\partial \alpha}Z_i(\boldsymbol{\beta})=\mathbf{0}, \tag{3.20}$$

In the next section we will see that there are certain advantages to basing estimation of α on minimization of an objective function.

3.2.1 Benefits of a Least Squares Approach for Estimation of α

One advantage to basing estimation of α on minimization of an objective function such as the generalized error sum of squares $Q^*(\alpha,\beta)$ is that this always results in one form of the estimator for α. This can lead to some simplification with regard to choice of estimator. For example, previously several different estimators have been suggested for the AR(1) correlation structure for GEE, in Liang and Zeger (1986), Newton (1988), and Wang and Carey (2003). In contrast, we will see that application of QLS leads to one estimator that is a function of the Pearson residuals.

A least squares approach for estimation of α is also helpful in avoiding some of the problems that had been noted in the literature with regard to the moment estimates for α that were implemented for GEE. Perhaps one of the earliest to note a drawback to the use of moment estimates was Park (1993), who noted that the GEE estimate of $R(\alpha)$ is not necessarily positive definite. Failure to achieve a positive-definite correlation matrix is serious because this will typically result in failure to converge for the iterative GEE estimation procedure.

Later, Crowder (1995) described the difficulties that can occur with respect to a moment-based estimating equations approach for estimation of the correlations, if the true structure is misspecified. He considered the simple situation in which the true and working correlation structures are among the AR(1) and exchangeable, but the working structure has been misspecified. First, if the AR(1) structure is misspecified as exchangeable, then it is easy to show that the explicit solution to a moment-based estimating equation for the exchangeable structure will not be consistent for the true correlation parameter. However, a well-established attractive feature of GEE is that $\widehat{\beta}_{GEE}$ will be consistent even if $\widehat{\alpha}_{GEE}$ is not consistent, as long as $\widehat{\alpha}_{GEE}$ converges to some value.

A potentially more serious situation occurs when the true exchangeable structure is misspecified as AR(1), in which case an estimate of the correlation parameter may fail to exist in the interval $(-1,1)$. Using the definitions of structures provided in Section 2.3.3 and the notation defined in Section 2.3.4, we consider the situation in which the *working* correlation structure $R_i(\alpha)$ is AR(1), but the *true* correlation structure $T_i(\rho)$ is exchangeable. We also consider the balanced data situation, for which n measurements were collected on each subject. For balanced data, the working and true structures will have the same dimensions for all subjects, so that we can drop the subscripts for this example, and refer to the working versus true structures as $R(\alpha)$ versus $T(\rho)$, respectively. In the discussions that follow, we shall see that it is helpful to distinguish between the "working" parameter α versus the "true" correlation parameter ρ when we consider the situation in which the true correlation structure has been misspecified.

Specification of a moment-based estimating equation for α is based on the expected value of the product of the jth and k^{th} Pearson residuals z_{ij} and z_{ik} on a subject, where $E(z_{ij}z_{ik}) = \phi Corr(y_{ij}, y_{ik})$. Because we assumed the correlations are AR(1), a moment-based estimating equation for α is as follows:

$$\sum_{i=1}^{m}\sum_{j<k}\left(\widehat{z}_{ij}\widehat{z}_{ik} - \phi\alpha^{|j-k|}\right) = 0, \tag{3.21}$$

where $\widehat{z}_{ij}$ is the Pearson residual that is evaluated at the current estimate of $\boldsymbol{\beta}$. Note also that $\sum_{i=1}^{m}\sum_{j<k}\phi\alpha^{|j-k|} = m\phi\frac{\alpha}{1-\alpha}\left(n - \frac{1-\alpha^n}{1-\alpha}\right)$, so that Equation (3.21) can be simplified as

$$\frac{1}{m}\sum_{i=1}^{m}\sum_{j<k}\widehat{z}_{ij}\widehat{z}_{ik} - \phi\frac{\alpha}{1-\alpha}\left(n - \frac{1-\alpha^n}{1-\alpha}\right) = 0. \tag{3.22}$$

The moment-based estimating equation approach for α will then alternate between updating $\widehat{\alpha}$ by obtaining a solution to Equation (3.22) for α, and then updating $\widehat{\boldsymbol{\beta}}$ by solving the GEE estimating equation for $\boldsymbol{\beta}$ evaluated at the current value of $\widehat{\alpha}$. However, Crowder (1995) pointed out that this iterative approach may break down because a solution to Equation (3.22) may fail to exist: Because the true correlation structure is exchangeable, $E(z_{ij}z_{ik}) = \phi\rho$. In addition, although the correlation structure has been misspecified, $\widehat{\boldsymbol{\beta}}$ is still consistent for $\boldsymbol{\beta}$. As a result, as $m \to \infty$, $\frac{1}{m}\sum_{i=1}^{m}\sum_{j<k}\widehat{z}_{ij}\widehat{z}_{ik}$ converges in probability to $\phi\frac{n(n-1)}{2}\rho$. The solution to Equation (3.22) therefore approaches the solution (for (α)) to the following estimating equation as $m \to \infty$:

$$\frac{n(n-1)}{2}\rho - \frac{\alpha}{1-\alpha}\left(n - \frac{1-\alpha^n}{1-\alpha}\right) = 0, \tag{3.23}$$

Crowder noted that Equation (3.23) has no real solution for $\alpha \in (-1,1)$ when n is odd and $-1/(n-1) \le \rho < -1/n$. For example, assume that $n = 3$ and $\rho = -0.49$. Then, we can simplify Equation (3.23) as

$$\alpha^4 - 1.53\alpha^2 - 0.94\alpha + 1.47 = 0, \tag{3.24}$$

which has no real solutions for $-1 < \alpha < 1$.

Because an iterative procedure based on moment-based estimating equations may break down for even the simple situation in which an exchangeable structure is misspecified as AR(1), Crowder suggested minimization of an objective function as one strategy that might ensure the existence of an estimate of α. This is the approach taken by QLS, which as we have seen, minimizes an objective function, the generalized error sum of squares, in stage one of the procedure. The early papers on QLS proved the existence of feasible estimators for α on a case by case basis, as we shall see demonstrated in this chapter for the AR(1) structure, and for other structures in subsequent chapters. Recently, Xie et al. (2010) proved that a feasible stage one estimator for α will *always* exist, while the stage two estimator will always be consistent for a particular class of correlation structures; we will discuss this in greater detail in Chapter 4.

A more fundamental objection: It could be argued that Crowder's objection regarding the potential lack of a feasible solution to a moment-based estimating equation for α has been largely overcome in practice, although the solution may not be completely satisfactory. For example, the estimating function for the AR(1) structure provided in Equation (3.21) is a nonlinear function of α based on the assumption that $E(z_{ij}z_{ik}) = \phi\alpha^{|j-k|}$, so that an iterative approach such as bisection would be required to obtain a solution in practice; Crowder had proven that solutions may fail to exist in certain situations, as demonstrated in Equation (3.24). However, current implementations of GEE largely avoid iterative approaches when updating the estimate of the correlation parameter.

For example, the approach employed in PROC GENMOD in SAS only makes use of information on the consecutive measurements of a subject, in order to obtain a *linear* estimating function of α. To see this, first consider the estimating equation

$$\sum_{i=1}^{m}\sum_{j=2}^{n_i}(z_{ij}z_{ij-1} - \phi\alpha) = 0, \tag{3.25}$$

which has the following solution for α:

$$\alpha = \frac{\sum_{i=1}^{m}\sum_{j=2}^{n_i}z_{ij}z_{ij-1}}{\sum_{i=1}^{m}(n_i - 1)\phi}. \tag{3.26}$$

Substitution of a consistent estimate for ϕ, $\widehat{z}_{ij}$ for z_{ij}, and a slight adjustment for the dimension of $\boldsymbol{\beta}$ in Equation (3.26) then yields the moment estimator $\widehat{\alpha}_{GEE-TRI}$ for the tri-diagonal structure that is implemented in PROC GENMOD for SAS and that was provided in Equation (2.19) in Section 2.3.5.

It is interesting to note that PROC GENMOD in SAS also implements $\widehat{\alpha}_{GEE-TRI}$ = $\widehat{\alpha}_{GEE-AR1}$ for the AR(1) structure. (See Equation (2.19) in Section 2.3.5.) This is because (3.26) is consistent for the true correlation parameter, if the true correlation structure is among the AR(1), tri-diagonal, and exchangeable. Why is this true? Estimating Equation (3.25) is an unbiased estimating equation as long as the off-diagonal elements of the true correlation structure are all equal, which is the case for the AR(1), tri-diagonal, and exchangeable structures. (See Exercise 3.6.) Solutions of unbiased estimating equations are typically consistent, so that $\widehat{\alpha}_{GEE-TRI}$ will be consistent for the true correlation parameter if the true structure is AR(1), tri-diagonal, or exchangeable.

This means that the estimator $\widehat{\alpha}_{GEE-AR1}$ defined in (2.19) will be consistent for α, even for Crowder's example that was discussed in Section 3.2.1; for this particular example, the moment-based estimating equation (3.21) does not have any feasible solutions when the true exchangeable structure is misspecified as AR(1), $n = 3$, and the true correlation parameter is -0.49. However, $\widehat{\alpha}_{GEE-AR1}$ ignores information on all but the consecutive Pearson residuals on each subject and could result in a loss in efficiency in estimation of α for small samples; it is perhaps for this reason that it is not implemented by PROC GENMOD for the exchangeable structure, even though it is a consistent estimator when the true exchangeable structure is misspecified as tri-diagonal, or AR(1). (See Problem 3.10.)

Crowder's complaint that feasible (positive definite) estimators may fail to exist for GEE can also be applied to the ML approach and to the *stage two* estimators of α for QLS, for some (but not all) correlation structures. For example, it was observed in Sabo and Chaganty (2009) that even for the simple situation in which an identity link function for a *correctly specified* correlation structure is implemented in analysis of correlated familial data, the moment-based (GEE), ML, and QLS estimators all yielded infeasible estimates of α in some simulation runs; however, the proportion of infeasible estimates was generally lower for QLS in contrast to ML and GEE.

However, in addition to his discussion of infeasibility and potential lack of consistency of the GEE estimator of α, Crowder had a deeper and potentially more serious objection to GEE. Crowder (1995) noted that if we stress that the correlation structure is unknown, which one might argue has been the tradition for GEE, then "α has no parametric identity independent of the particular estimating functions chosen; α does not exist in any fundamental sense." We suspect that the tendency in the GEE literature to downplay the importance of correct selection of the correlation structure may be due to the fact that there was some discomfort on the part of the original authors of GEE with respect to their proposed estimators for α. For example, as mentioned in Section 2.3.5 of the previous chapter, Liang and Zeger (1986) proposed a plotting technique for estimation of α for the AR(1) structure that is problematic, as discussed in Shults and Chaganty (1998) and reflected in the fact that it has not been applied in any of the major software packages that implement GEE.

We do acknowledge that a valid reason for treating the modeling and estimation of α to be of secondary importance to careful modeling of $\boldsymbol{\beta}$, is that even when the true correlation structure is misspecified, GEE will typically yield consistent estimates of the regression parameter, albeit with a potential loss in efficiency in estimation of the regression parameter. However, we agree with Crowder that the role of α is unclear, if we stress that we do not know the true working correlation structure. We may refer to the specified structure as a "working structure" because this has been the tradition in the GEE literature. However, we could just as easily have referred to our choice of link function, or any other assumption in our model, as a "working assumption." In other words, even though the estimation procedure for GEE may be more robust to misidentification of the correlation structure than it is to misidentification of the model for the marginal mean, we will proceed with estimation and testing under the assumption that all our model assumptions are correct; we will then explore the impact of violation of our model assumptions, for example, the sensitivity of our results to misidentification of the correlation structure, or failure to include an important covariate in the model for the marginal mean. We will not take the view that we "can never know" the correlation structure because as Crowder pointed out, in that case, α "fails to exist in any fundamental sense."

3.2.2   QLS Stage One Estimates of α for the AR(1) Structure

Here we demonstrate implementation of stage one of QLS for the AR(1) structure that is based on minimization of an objective function, the generalized error sum of squares. First, it is helpful to note the well-known result that the $n_i \times n_i$ AR(1) structure $R_i(\alpha)$ has an inverse that can be expressed in the following form:

$$R_i^{-1}(\alpha) = \frac{1}{1-\alpha^2}\left(I_{n_i} + \alpha^2 C_{2i} - \alpha C_{1i}\right), \qquad (3.27)$$

where C_{1i} and C_{2i} have dimension $n_i \times n_i$; C_{1i} is a tri-diagonal matrix with ones on the upper and lower diagonal matrix; and $C_{2i} = diag(0,1,...,1,0)$.

The derivative of $R_i^{-1}(\alpha)$ can then be directly obtained as

$$\frac{\partial R_i^{-1}(\alpha)}{\partial \alpha} = \frac{-1}{(1-\alpha^2)^2}\left(\alpha^2 C_{1i} - 2\alpha\left[I_{n_i} + C_{2i}\right] + C_{1i}\right). \qquad (3.28)$$

Substitution of Equation (3.28) into the left-hand side of Equation (3.17) then yields the following QLS stage one estimating equation for the AR(1) structure:

$$\alpha^2 \sum_{i=1}^{m} Z_i' C_{1i} Z_i - 2\alpha\left[\sum_{i=1}^{m} Z_i' Z_i + \sum_{i=1}^{m} Z_i' C_{2i} Z_i\right] + \sum_{i=1}^{m} Z_i' C_{1i} = 0, \qquad (3.29)$$

which has solutions

$$\alpha = \frac{\sum_{i=1}^{m}\left(Z_i' Z_i + Z_i' C_{2i} Z_i\right) \pm \sqrt{\left(\sum_{i=1}^{m}\left(Z_i' Z_i + Z_i' C_{2i} Z_i\right)\right)^2 - \left(\sum_{i=1}^{m} Z_i' C_{1i} Z_i\right)^2}}{2\sum_{i=1}^{m} Z_i' C_{1i} Z_i}. \qquad (3.30)$$

Shults (1996) noted that the solutions Equation (3.30) will be real if and only if

$$\left(\sum_{i=1}^{m}\left(Z_i' Z_i + Z_i' C_{2i} Z_i\right)\right)^2 - \left(\sum_{i=1}^{m} Z_i' C_{1i} Z_i\right)^2 \geq 0 \iff$$

$$\left(\sum_{i=1}^{m}\left(Z_i' Z_i + Z_i' C_{2i} Z_i\right) + \sum_{i=1}^{m} Z_i' C_{1i} Z_i\right)\left(\sum_{i=1}^{m}\left(Z_i' Z_i + Z_i' C_{2i} Z_i\right) - \sum_{i=1}^{m} Z_i' C_{1i} Z_i\right) \geq 0 \iff$$

$$\sum_{i=1}^{m}\sum_{j=1}^{n_i-1}\left(z_{ij} + z_{ij+1}\right)^2 \sum_{i=1}^{m}\sum_{j=1}^{n_i-1}\left(z_{ij} - z_{ij+1}\right)^2 \geq 0,$$

which clearly is true. For the special case of balanced data, Chaganty (1997) used the Sturm sequence property of tri-diagonal matrices to prove the solutions to Equation (3.30) will be real, in a proof that set up Kronecker product matrices and required the AR(1) correlation matrices $R_i(\alpha)$ to be of equal dimension.

Next, simple arithmetic can be used to show that the following stage one estimator for the AR(1) structure (obtained as the solution (3.30) with a negative sign in the numerator) will always take value in the interval $(-1,1)$:

$$\widehat{\alpha}_{QONE} = \frac{\sum_{i=1}^{m}\sum_{j=2}^{n_i}\left(z_{ij}^2 + z_{ij-1}^2\right) - \sqrt{\sum_{i=1}^{m}\sum_{j=2}^{n_i}\left(z_{ij}^2 + z_{ij-1}^2\right)\sum_{i=1}^{m}\sum_{j=2}^{n_i}\left(z_{ij}^2 - z_{ij-1}^2\right)}}{2\sum_{i=1}^{m}\sum_{j=2}^{n_i} 2z_{ij}z_{ij-1}}. \qquad (3.31)$$

Because the stage one estimator (3.31) will always take value in the feasible (positive definite) interval $(-1, 1)$ for the AR(1) structure, the estimated correlation matrix will be positive definite for all the sample sizes. The stage one QLS estimator $\widehat{\alpha}_{QONE}$ is a function of the Pearson residuals z_{ij}, which will be true for other correlation structures as well. When we are unable to obtain an explicit solution to the stage one estimating function (3.17) for α, we will see in subsequent chapters that we can use the bisection method or some other iterative approach to obtain a solution $\widehat{\alpha}_{QONE}$.

3.2.3   Limiting Value of the Stage One QLS Estimator of α

The stage one estimate $\widehat{\alpha}_{QONE}$ for the AR(1) structure will exist and will take a value in the feasible interval $(-1, 1)$ with probability 1. However, there is a serious drawback: $\widehat{\alpha}_{QONE}$ is not consistent, even when the working correlation structures have been correctly specified, so that in the notation of Section 2.3.4, the working structures $R_i(\alpha)$ are identical to the true structures $T_i(\rho)$ and $\rho = \alpha$.

As in Chaganty and Shults (1999), our strategy to prove that $\widehat{\alpha}_{QONE}$ is not consistent will be to first note that $\widehat{\alpha}_{QONE}$ is the solution (for α) to the stage one estimating equation (3.17), so that $\widehat{\alpha}_{QONE}$ satisfies Equation (3.19). Next, we will show that the limiting value of the solution (for $\widehat{\alpha}_{QONE}$) to Equation (3.20) is not equal to α, but instead is equal to some function $g(\alpha)$ of α. A second stage of the procedure will therefore be necessary in order to make a correction to the stage one estimator that yields a consistent estimator of α.

To prove the results that we just stated above, we start by noting that the dimension of quadratic form $Z_i'(\boldsymbol{\beta})\frac{\partial R_i^{-1}(\alpha)}{\partial \alpha}Z_i(\boldsymbol{\beta})$ is 1×1, so that $Z_i'(\boldsymbol{\beta})\frac{\partial R_i^{-1}(\alpha)}{\partial \alpha}Z_i(\boldsymbol{\beta})$ is equal to its trace, that is,

$$Z_i'(\boldsymbol{\beta})\frac{\partial R_i^{-1}(\alpha)}{\partial \alpha}Z_i(\boldsymbol{\beta}) = trace\left(Z_i'(\boldsymbol{\beta})\frac{\partial R_i^{-1}(\alpha)}{\partial \alpha}Z_i(\boldsymbol{\beta})\right).$$

In addition, note that the expected value of $Z_i(\boldsymbol{\beta})Z_i'(\boldsymbol{\beta})$ is equal to ϕ times the correlation matrix of Y_i, so that

$$E\left(Z_i(\boldsymbol{\beta})Z_i'(\boldsymbol{\beta})\right) = \phi R_i(\alpha).$$

Next, it is helpful to recall that $trace(AB) = trace(BA)$. We then see that

$$
\begin{aligned}
E\left(Z_i'(\boldsymbol{\beta})\frac{\partial R_i^{-1}(\alpha)}{\partial \alpha}Z_i(\boldsymbol{\beta})\right) &= trace\left(\frac{\partial R_i^{-1}(\alpha)}{\partial \alpha}E\left(Z_i(\boldsymbol{\beta})Z_i'(\boldsymbol{\beta})\right)\right) \\
&= \phi trace\left(\frac{\partial R_i^{-1}(\alpha)}{\partial \alpha}R_i(\alpha)\right).
\end{aligned}
$$

It then follows that asymptotically the solution $\widehat{\alpha}_{QONE}$ (for α) to estimating Equation (3.20) will satisfy the following estimating equation

$$\sum_{i=1}^{m} trace\left(\frac{\partial}{\partial \alpha}R_i^{-1}(\alpha)\bigg|_{\alpha=\widehat{\alpha}_{QONE}} R_i(\alpha)\right) = 0 \tag{3.32}$$

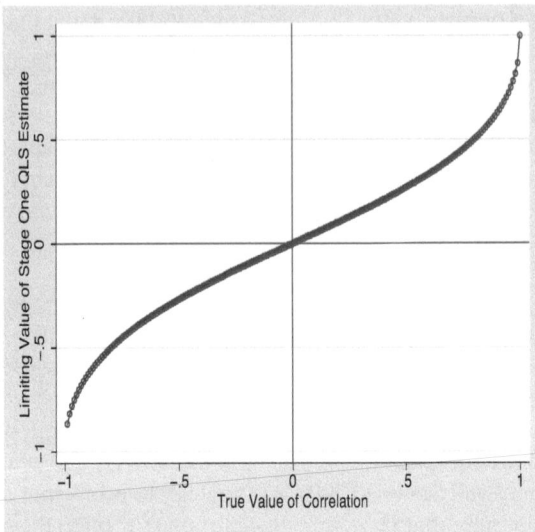

Figure 3.1 *Limiting value of stage one QLS estimate* $\widehat{\alpha}_{QONE}$ *versus* α *when the AR(1) struc-ture has been correctly specified.*

as $m \longrightarrow \infty$; thus, $\widehat{\alpha}_{QONE}$ converges in probability to the solution (for $\widehat{\alpha}_{QONE}$) to (3.32) as $m \longrightarrow \infty$.

When the correlation structure is AR(1), as noted in Chaganty and Shults (1999), the solution (for $\widehat{\alpha}_{QONE}$) to Equation (3.32) is

$$\widehat{\alpha}_{QONE} = g(\alpha) = \frac{1 - \sqrt{1 - \alpha^2}}{\alpha} \tag{3.33}$$

for $\alpha \neq 0$, or 0 for $\alpha = 0$. Therefore, we see that

$$\widehat{\alpha}_{QONE} \xrightarrow{p} g(\alpha), \tag{3.34}$$

where $g(\alpha)$ is provided in Equation (3.33).

Figure 3.1 displays the limiting value of $\widehat{\alpha}_{QONE}$ versus α and a 45-degree line that represents the limiting value of a consistent estimator of α. As shown in Figure 3.1, the asymptotic bias is negative for $\alpha > 0$ and positive for $\alpha < 0$. The asymptotic bias is also potentially severe, e.g. if $\alpha = 0.90$ or 0.95 then the limiting values are 0.63 and 0.72, respectively. We therefore require a correction to the stage one estimate, in order to achieve consistency in estimation of α.

3.3 Stage Two QLS Estimates of the Correlation Parameter for the AR(1) Structure

3.3.1 Elimination of the Asymptotic Bias in the Stage One QLS Estimate of α

Here we continue our demonstration of QLS for the AR(1) structure, in order to pro-vide insight into both the need for a second stage of the QLS procedure and a general approach that we might take in order to achieve consistency in estimation of α: Because the stage one estimate $\widehat{\alpha}_{QONE}$ is asymptotically biased, Chaganty and Shults (1999) introduced a second stage to the QLS procedure that utilizes the limiting value of $\widehat{\alpha}_{QONE}$ to obtain a consistent estimate $\widehat{\alpha}_{QLS}$ of α. First, note that

$$\widehat{\alpha}_{QONE} \xrightarrow{p} g(\alpha) \implies \tag{3.35}$$

$$g^{-1}(\widehat{\alpha}_{QONE}) \xrightarrow{p} \alpha, \tag{3.36}$$

where $g(\alpha)$ is provided in Equation (3.33) and can be easily to shown to have inverse $g^{-1}(\alpha) = 2\alpha/(1+\alpha^2)$.

Equation (3.35) therefore tells us that we can make a correction to the stage one estimator $\widehat{\alpha}_{QONE}$ in order to obtain a consistent estimator $\widehat{\alpha}_{QLS}$ of α that is defined as follows:

$$\widehat{\alpha}_{QLS} = \frac{2\widehat{\alpha}_{QONE}}{1+\widehat{\alpha}_{QONE}^2} \tag{3.37}$$

for the AR(1) structure and for $\alpha \neq 0$, where $\widehat{\alpha}_{QONE}$ is provided in Equation (3.31). As a result, the stage two estimates $\widehat{\alpha}_{QLS}$ will converge in probability to the true value of α. Figure 3.2 updates Figure 3.1 to display the limiting values versus α of the stage one QLS estimate $\widehat{\alpha}_{QONE}$ (on the left) and of the stage two estimates $\widehat{\alpha}_{QLS}$ (on the right), when the working correlation structure has been correctly specified as AR(1). Figure 3.2 displays visually that the limiting values are not the true values for stage one, but that the estimators do tend to the correct values for stage two.

A very important result for the AR(1) structure is that *the stage two estimator* $\widehat{\alpha}_{QLS} = g^{-1}(\widehat{\alpha}_{QONE})$ *will take value in the positive-definite (feasible) interval* $(-1, 1)$ *for any sample size*, so that the estimated correlation matrix will always be positive definite for the AR(1) correlation structure. This is easily proven as follows. First, we had proven in Section 3.2.2 that the stage one estimator $\widehat{\alpha}_{QONE}$ will always take value in $(-1, 1)$. Next, note that $g^{-1}(-1) = -1$; $g^{-1}(1) = 1$; and $\frac{\partial}{\partial \alpha} g^{-1}(\alpha) = 2(1 - \alpha^2)/(1+\alpha^2)^2$, which is positive for $\alpha \in (-1, 1)$; thus $g^{-1}(\alpha)$ is strictly increasing over the interval $(-1, 1)$. It therefore follows that $\widehat{\alpha}_{QLS} = g^{-1}(\widehat{\alpha}_{QONE})$ takes value in $(-1, 1)$ for $\widehat{\alpha}_{QONE} \in (-1, 1)$, so that the stage two estimator of α is guaranteed to be feasible (positive definite) for the AR(1) correlation structure.

More generally, note that $\widehat{\alpha}_{QLS}$ can also be obtained as a solution to estimating equation (3.32) for α. As in the AR(1) case, the second stage of QLS therefore applies a correction to the stage one estimate $\widehat{\alpha}_{QONE}$ by calculating $g^{-1}(\widehat{\alpha}_{QONE})$, where $g(\alpha)$ is the limiting value of the stage one estimate $\widehat{\alpha}_{QONE}$. While the stage one estimate $\widehat{\alpha}_{QONE}$ is a function of the Pearson residuals, the stage two estimate will be a function of $\widehat{\alpha}_{QONE}$ only. The function $g(\cdot)$ will vary according to the choice of working correlation structure.

If we specify the wrong working correlation structures, this will often result in application of the wrong correction to $\widehat{\alpha}_{QONE}$ when we obtain the stage two estimate of α. As we shall see, in these situations we can easily obtain the limiting value of

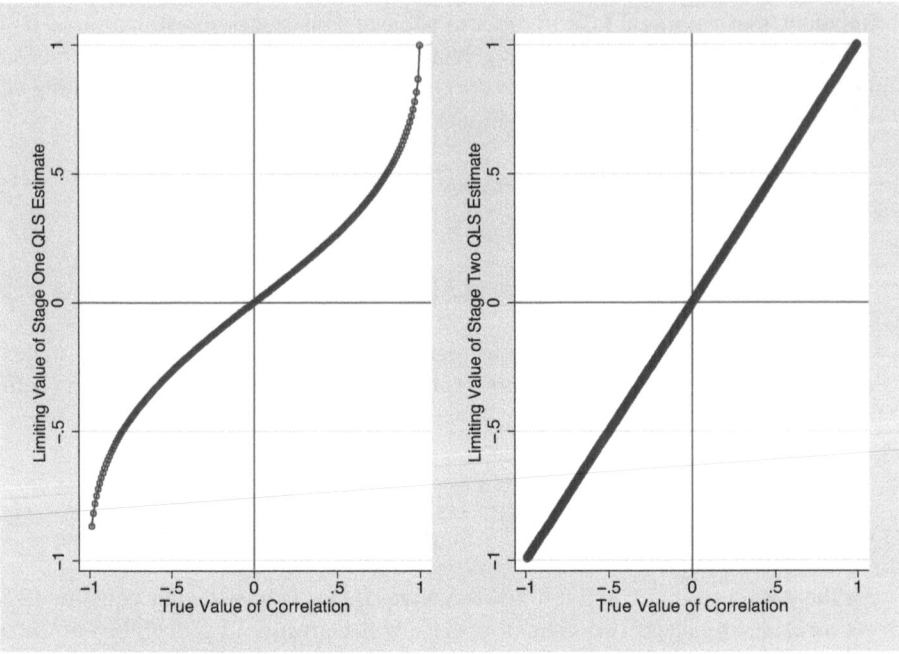

Figure 3.2 *(LEFT) Limiting value of stage one QLS estimate $\widehat{\alpha}_{QONE}$ versus α when the AR(1) structure has been correctly specified. (RIGHT) Limiting value of stage two QLS estimate $\widehat{\alpha}_{QLS}$ versus α when the AR(1) structure has been correctly specified.*

$\widehat{\alpha}_{QONE}$. For example, if the true structure required a correction of $f^{-1}(\widehat{\alpha}_{QONE})$ but we incorrectly assumed the AR(1) structure and applied a correction of $g^{-1}(\widehat{\alpha}_{QONE})$, then $\widehat{\alpha}_{QONE}$ would converge in probability to $g^{-1}(f(\alpha))$, which will not necessarily equal α.

However, it is important to note the following: As discussed in Chaganty and Shults (1999), $g^{-1}(\widehat{\alpha}_{QONE})$ will be consistent for α *even if the true correlation struc-ture is not AR(1)*, as long as the off-diagonal elements of the true correlation structure are equal, that is, as long as the correlation is constant between consecutive observations on a subject. This follows from the fact that the inverse of the AR(1) correlation structure is tri-diagonal: Suppose we have incorrectly specified the true correlation structures $T_i(\rho)$ as AR(1) structures $R_i(\alpha)$, so that the true structures $T_i(\rho)$ are not identical to the working structures $R_i(\alpha)$. Then, using the same arguments that were given at the start of the previous section, we can see that the solution to the stage one estimating equation (3.17) will converge in probability to the solution (for α) to the following estimating equation:

$$trace\left(\sum_{i=1}^{m}\frac{\partial}{\partial\alpha}R_i^{-1}(\alpha)T_i(\rho)\right)=0. \qquad (3.38)$$

Because $\frac{\partial}{\partial\alpha}R_i^{-1}(\alpha)$ is a tri-diagonal matrix for the AR(1) structure, the left-hand side of Equation (3.38) only involves the off-diagonal elements of $T_i(\rho)$. This means that the left-hand side of (3.38) will be the same when $T_i(\rho)$ is exchangeable or tri-diagonal, as it was when $T_i(\rho)$ was AR(1). As a result, the solution (for α) to Equation (3.38) will be the same for the AR(1), tri-diagonal, and exchangeable structures. (See Exercise 3.10.) In general, the same correction to achieve consistency for the stage two QLS estimator for the AR(1) working structure will be required, as long as the off-diagonal elements of the true correlation matrices are equal. This is a helpful result because it ensures, for example, that $\widehat{\alpha}_{QLS}$ will be consistent for the true parameter ρ, if the working structure is AR(1) and the true structure is among the AR(1), tri-diagonal, or exchangeable; however, note that we can suffer a loss in efficiency in estimation of $\boldsymbol{\beta}$ if the true correlation structure is misspecified.

An additional benefit of adding a second stage to the procedure: In addition to providing a consistent and feasible estimate, Sun et al. (2009) compared the stage one and stage two QLS estimators and demonstrated that the second stage of QLS offered a great improvement over stage one in terms of reduced mean square error (MSE) for higher values of α. They explained the superiority of the two-stage versus one-stage QLS approach simulation results by noting that the MSE of $\widehat{\alpha}_{QONE}$ can be reduced by use of the stage two estimate $\widehat{\alpha}_{QLS}$ in two ways: First, consider the decomposition formula for the MSE of an estimator, $\text{MSE}(\widehat{\alpha}_{QLS}) = [\text{Bias}(\widehat{\alpha}_{QLS})]^2 + \text{Var}(\widehat{\alpha}_{QLS})$. Next, note that the first term in the MSE is reduced by use of the stage two estimate $\widehat{\alpha}_{QLS}$ versus the stage one estimate $\widehat{\alpha}_{QONE}$ because $\widehat{\alpha}_{QONE}$ is biased, while $\widehat{\alpha}_{QLS}$ is consistent. In addition, the variance of $\widehat{\alpha}_{QLS}$ might be reduced by the function that relates the stage one and stage two estimates. For example, as we saw, the function for the AR(1) structure is given by $\widehat{\alpha}_{QLS} = g^{-1}(\widehat{\alpha}_{QONE}) = 2\widehat{\alpha}_{QONE}/(1 + \widehat{\alpha}_{QONE}^2)$.

According to the Delta method, the asymptotic variance of the stage two estimator satisfies

$$\text{AsyVar}(\widehat{\alpha}_{QLS}) = \text{dginv}(\widehat{\alpha}_{QLS})\text{Var}(\widehat{\alpha}_{QONE})\text{dginv}(\widehat{\alpha}_{QLS}),$$

where

$$\text{dginv}(\alpha) = \frac{\partial g^{-1}(\alpha)}{\partial \alpha} = 2(1 - \alpha^2)/(1 + \alpha^2)^2.$$

The second term of the MSE is therefore also reduced when $|\alpha| > 0.49$; for example, when $\alpha = 0.80$, $\text{AsyVar}(\widehat{\alpha}_{QLS}) \approx 0.07\text{Var}(\widehat{\alpha}_{QONE})$. However, for smaller α it is interesting to note that the second term of the MSE is not reduced; for example, when $\alpha = 0$, $\text{AsyVar}(\widehat{\alpha}_{QLS}) = 2\text{Var}(\widehat{\alpha}_{QONE})$.

3.4 Algorithm for QLS

The same approach taken to provide a consistent estimator for α for the AR(1) structure can be applied for other structures. For the AR(1) structure, we were able to obtain explicit solutions to the QLS stage one and stage two estimating equations for the correlation parameter. For other structures, explicit solutions may not be available, in which case the method of bisection or another iterative approach might be applied. We will provide details for particular structures in subsequent chapters. We can combine the two stages of QLS to yield the following procedure for estimation of α and β:

Algorithm for Estimation of the Regression and Correlation Parameters with QLS:

Stage One of QLS: First, select a starting value for the stage one estimate $\widehat{\alpha}_{QONE}$, e.g. $\widehat{\alpha}_{QONE} = 0$. Next, alternate between Steps A and B:

Step A : Update the estimate of regression parameter β by solving the GEE estimating equation for β:

$$\sum_{i=1}^{m} D_i' A_i^{-1/2} R_i^{-1} (\widehat{\alpha}_{QONE}) A_i^{-1/2} (Y_i - U_i(\beta)) = 0; \tag{3.39}$$

and

Step B : Update the estimate of the correlation by obtaining a solution (for α) to the stage one estimating equation evaluated at the current estimate $\widehat{\beta}$ for β:

$$\sum_{i=1}^{m} Z_i'(\widehat{\beta}) \frac{\partial R_i^{-1}(\alpha)}{\partial \alpha} Z_i(\widehat{\beta}) = 0. \tag{3.40}$$

Stage Two of QLS: After convergence in the parameters in stage one, we next obtain a consistent estimate of α by solving the stage two estimating equation for α:

$$\sum_{i=1}^{m} trace \left(\frac{\partial R_i^{-1}(\alpha)}{\partial \alpha} \bigg|_{\alpha = \widehat{\alpha}_{QONE}} R_i(\alpha) \right) = 0. \tag{3.41}$$

Theorem 3.2 of Chaganty and Shults (1999) established that if there is a unique solution $\alpha = g^{-1}(\widehat{\alpha}_{QONE})$ to Equation (3.41) that is a one to one and continuous function of $\widehat{\alpha}_{QONE}$ and the structure is correctly specified, then the stage two estimate $\widehat{\alpha}_{QLS} = g^{-1}(\widehat{\alpha}_{QONE})$ is consistent for α. The final QLS estimate of β is then obtained by solving the GEE estimating equation (3.39) (evaluated at $\widehat{\alpha}_{QLS}$) for β.

Note that estimation of the scalar parameter ϕ is not involved in the iterative estimation procedure for QLS. After obtaining the QLS estimates of α and of β, the QLS estimate of the scalar parameter ϕ can be obtained as follows: $\widehat{\phi} = \min\left\{ \widehat{\phi}_p, \widehat{\phi}_c \right\}$, for

$$\widehat{\phi}_p = \frac{1}{m} \sum_{i=1}^{m} \frac{Z_i(\widehat{\beta})' Z_i(\widehat{\beta})}{n_i} \text{ and } \widehat{\phi}_c = \frac{1}{m} \sum_{i=1}^{m} \frac{Z_i(\widehat{\beta})' R_i^{-1}(\widehat{\alpha}_{QLS}) Z_i(\widehat{\beta})}{n_i}.$$

Another view of QLS as an unbiased estimating equations approach: As discussed in Remark 3.3 of Chaganty and Shults (1999), QLS provides solutions to an unbiased estimating equation that does not involve estimation of the scalar parameter ϕ. Sun et al. (2009) provided the following summary. They referred to the estimating function on the left-hand side of Equation (3.40), which represents the first derivative of the generalized error sum of squares with respect to α, as $D_G(\alpha)$. Next, note that QLS obtains the stage one estimate for α by solving $D_G(\alpha) = 0$ for α. QLS then obtains the expectation of $D_G(\alpha)$ and obtains the stage two estimator of the true correlation parameter by solving $E(D_G(\alpha)) = 0$ for α. QLS therefore employs a two-stage procedure to obtain a solution to the unbiased estimating equation $D_G(\alpha) - E(D_G(\alpha)) = 0$ for α.

3.4.1   Asymptotic Distribution of the Regression Parameter for QLS

The asymptotic distribution of the QLS estimate $\widehat{\boldsymbol{\beta}}_{QLS}$ is the same as the asymptotic distribution of the GEE estimate $\widehat{\boldsymbol{\beta}}_{GEE}$, as was proven in Chaganty (1997). Therefore, the asymptotic covariance matrix of $\widehat{\boldsymbol{\beta}}_{GEE}$ that was provided in Equation (2.23) applies for both QLS and GEE.

3.5   Other Approaches Based on GEE

There are other extensions of GEE that, in contrast to GEE, do not treat the correlations as nuisance parameters. These approaches involve models for correlation structures often leading to more complex weighting matrices than the original formulation of GEE (sometimes referred to as first-order GEE). Prentice (1988) developed the GEE1 approach for analysis of correlated binary outcomes by constructing a second estimating equation (Equation (14) on p. 1039 of Prentice (1988)) that involves specifying correlation structures for the sample correlations. Zhao and Prentice (1990) and Liang et al. (1992) developed GEE2 (second-order GEE) approaches for the analysis of discrete and continuous outcomes; they accomplished this by constructing a joint estimating equation for the marginal mean regression and correlation regression parameters that involves a covariance structure that depends on the third and fourth moments of the outcome variable. An important difference between GEE and GEE2 is that GEE2 does not treat the correlation and regression parameter vectors as though they are orthogonal to one another. As a result GEE2 is less robust to misidentification of the correlation structure than is GEE. However, when the number of subjects is small or the correlations are of primary interest, then GEE2 may be preferable to GEE, assuming of course that the correlation structure has been correctly specified. See Section 3.4 of Hardin and Hilbe (2003) for a more detailed description of GEE2. Later approaches based on GEE1 and GEE2, for example methods for hierarchical data by Qu et al. (1992) and Qaqish and Liang (1992), also involve specification of complex weighting structures.

Carey et al. (1993) developed alternating logistic regressions (ALR) for the marginal mean regression of correlated binary outcomes by estimating within-cluster association among the outcomes with a GEE based on conditional means (Equation (7) on p. 521) that corresponds to the logistic regression of one response on another. Their approach also involves specification of a weighting matrix that depends on higher-order moments of the outcome variable. An important distinction between ALR and GEE is that ALR models associations via odds-ratios, which are subject to less restrictive constraints than are the correlations α for binary data. Preisser et al. (2003a) implemented ALR in the analysis of irregularly spaced longitudinal binary data. Qaqish et al. (2012) generalize ALR in a procedure known as orthogonalized residuals. However, although the odds-ratios are not as constrained as are correlations, Guerra et al. (2012) demonstrated in extensive simulations that there can be a severe deterioration in performance for ALR for larger odds-ratios.

Each of the extensions of GEE described above potentially involves more complex patterned correlation structures than GEE, and they differ from GEE's usual implementation based on moment estimation of α by instead defining estimating equations for the correlation parameters defining those structures, with empirical sandwich variance estimation of $\hat{\beta}$ and $\hat{\alpha}$ as a byproduct. Because efficient estimation of α requires the introduction of parameters relating to moments higher than order 2, the typical approach in implementing each of these methods is to simplify the weighting structure in the estimating equations for α by setting some or all of the their off-diagonal elements equal to 0. For example, Carey et al. (1993) define their weighting matrix ((7), p. 521, 1993) as a diagonal matrix and find, in specific examples involving small cluster sizes, that this results in near-optimal weighting. When GEE1 and GEE2 are applied in analyses, off-diagonal elements of the weighting matrices are often set equal to equal to 0. However, it should be noted that the increased complexity of these estimation procedures for α coincides with improved flexibility in modeling within-clusters association structures.

It is also interesting that although many extensions of GEE have been proposed and continue to be discussed in the statistical literature, few of these methods have been implemented in available software. Some exceptions include ALR, which was implemented in the SAS procedure PROC GENMOD and in the ORTH package for R software that was developed by By et al. (2011). In addition, Qaqish and Preisser developed an SAS/IML macro for application of ALR and orthonalized residuals; the software also provides cluster-level deletion diagnostics (Preisser et al., 2012) for both "the marginal (logistic) model and the within-cluster association (log odds ratio) model":

http://www.bios.unc.edu/~preisser/personal/software.html.

In our research, we prefer to focus on methods that can be implemented using existing software or with programs provided by those who developed the approach. Even if we need to rewrite some programs, for example, to ensure a fair comparison between methods in a simulation study, the fact that the developers of the method provided software is a good sign that they believe in the usefulness of their approach. For more discussion of the many extensions of GEE that have been discussed in the literature, see the texts by Hardin and Hilbe (2012), Diggle et al. (2002), and Ziegler (2011).

3.6   Example

Here we demonstrate implementation of QLS for analysis of the longitudinal study in obesity described in Section 1.5.1 in Chapter 1 and used to demonstrate application of GEE in Section 2.4 in Chapter 2. Using the **xtqls** procedure (Shults et al., 2007) in Stata 12.0, we fit a regression model with an identity link function to regress BMI z-score on *month* and *month*-squared, while adjusting for the correlation between repeated measurements on a subject via application of an AR(1) working correlation structure.

We first fit QLS with a model-based covariance matrix using the **xtqls** procedure: (Note that **xtqls** must be installed prior to running these commands.)

Example   61

```
. use random_small, clear
. gen change = bmiz - basebmiz
. gen month2 = month^2
. keep if month<=36
(65 observations deleted)

. xtqls change month month2, i(id) c(AR 1) vce(model) f(gau) t(month)

Iteration 1: tolerance = .01733588
Iteration 2: tolerance = 0
```

```
GEE population-averaged model              Number of obs      =       466
Group and time vars:           id __00000S  Number of groups   =       100
Link:                             identity   Obs per group: min =         2
Family:                           Gaussian                  avg =       4.7
Correlation:            fixed (specified)                   max =         6
                                             Wald chi2(2)      =      2.10
Scale parameter:                 .8225255    Prob > chi2       =    0.3493
```

```
------------------------------------------------------------------------------
     change |     Coef.    Std. Err.      z    P>|z|     [95% Conf. Interval]
------------+-----------------------------------------------------------------
      month |   .0030446   .0086513     0.35   0.725    -.0139117    .0200009
     month2 |  -.0001788   .0002155    -0.83   0.406    -.0006011    .0002434
      _cons |   .8373935   .0920446     9.10   0.000     .6569894    1.017798
------------------------------------------------------------------------------
```

Next, we obtain the estimated correlation matrix.

```
. xtcorr

Estimated within-id correlation matrix R:

          c1      c2      c3      c4      c5      c6
r1   1.0000
r2   0.8275  1.0000
r3   0.6847  0.8275  1.0000
r4   0.5666  0.6847  0.8275  1.0000
r5   0.4689  0.5666  0.6847  0.8275  1.0000
r6   0.3880  0.4689  0.5666  0.6847  0.8275  1.0000
```

Next, we fit the same model with the sandwich-based covariance matrix.

```
. xtqls change month month2, i(id) c(AR 1) vce(robust) f(gau) t(month)

Iteration 1: tolerance = .01733588
Iteration 2: tolerance = 0
```

```
GEE population-averaged model              Number of obs      =       466
Group and time vars:           id __00000S  Number of groups   =       100
Link:                             identity   Obs per group: min =         2
Family:                           Gaussian                  avg =       4.7
Correlation:            fixed (specified)                   max =         6
                                             Wald chi2(2)      =      2.89
Scale parameter:                 .8225255    Prob > chi2       =    0.2353

                                   (Std. Err. adjusted for clustering on id)
```

```
----------------------------------------------------------------------
              |                 Semirobust
    change    |   Coef.      Std. Err.     z     P>|z|    [95% Conf. Interval]
--------------+-------------------------------------------------------------
     month    |  .0030446    .009685     0.31   0.753   -.0159377    .0220269
    month2    | -.0001788    .0002311   -0.77   0.439   -.0006317    .000274
     _cons    |  .8373935    .072301    11.58   0.000    .6956861    .9791009
----------------------------------------------------------------------
```

Note that the results obtained above were almost identical as to the analysis results for GEE that were provided at the end of Chapter 2.

As for the GEE analysis, *month* and *month*-squared did not differ significantly from zero for both the model-based and "robust" sandwich covariance matrix. In Chapter 4 we will implement a correlation structure that allows the correlation to depend on the actual separation in timing of measurements with QLS; for this analysis, the time effects will be significant, which will be consistent with our prior knowledge that children tend to have an initial weight gain following kidney transplant that is followed by a subsequent leveling off in weight.

3.7   Summary

QLS was developed over a period of several years, in several manuscripts. The QLS approach for estimation of the correlation parameters was motivated by the exploration of the link between LS, GLM, and GEE, in Dunlop (1994). While Dunlop had considered the error sum of squares Q^* as a potential objective function for least squares estimation of β, QLS took a least squares approach to estimation of the correlation parameter α, by minimizing the generalized error sum of squares $Q^*(\alpha, \beta)$ with respect to α in stage one of the QLS estimation procedure. QLS was named a "quasi"-least squares approach because although it takes an LS approach with respect to estimation of α in stage one of the procedure, it does not fully differentiate $Q^*(\alpha, \beta)$ with respect to β; as a result, the QLS estimates do not fully conform to the principle of least squares. We considered the AR(1) structure for the purpose of demonstration and obtained an explicit expression for the stage one estimate $\widehat{\alpha}_{QONE}$ as a function of the Pearson residuals for this structure. However, $\widehat{\alpha}_{QONE}$ is not consistent and we explained how to obtain a stage two estimate $\widehat{\alpha}_{QLS}$ that will be a function of $\widehat{\alpha}_{QONE}$, so that $\widehat{\alpha}_{QLS} = g^{-1}(\widehat{\alpha}_{QONE})$. Although we were able to obtain explicit expressions for the AR(1) structure, iterative approaches such as bisection might be required for other structures, to solve the stage one and stage two estimating equations to obtain $\widehat{\alpha}_{QONE}$ and $\widehat{\alpha}_{QLS}$, respectively. Overall, the estimation procedure for QLS can be described as a two-stage procedure for estimation of α that is in the framework of GEE because it alternates between updating the estimate of β (via solution of the GEE estimating equation for β) and updating the estimate of α (via solution of the stage one estimating equation for α in stage one, or solution of the stage two estimating equation in stage two of the procedure).

3.8   Exercises

Exercise 3.1 *Show that simplification of Equation (3.5) yields the normal equations* $X'X\boldsymbol{\beta} = X'\boldsymbol{Y}$.

Exercise 3.2 *Show that the first term of the estimating function in Equation (3.16) is the GEE estimating function for* $\boldsymbol{\beta}$.

Exercise 3.3 *Show that the QLS stage one and stage two estimates yield solutions to an unbiased estimating equation.*

Exercise 3.4 *Show that* $\sum_{i=1}^{m}\sum_{j<k}\phi\alpha^{|j-k|} = m\phi\frac{\alpha}{1-\alpha}\left(n - \frac{1-\alpha^n}{1-\alpha}\right)$, *so that Equation (3.21) can be simplified as Equation (3.22).*

Exercise 3.5 *Assume that* $n = 3$ *and* $\rho = -0.49$. *Verify that Equation (3.23) can then be simplified as Equation (3.24). Next, show that Equation (3.24) has no real solutions for* $-1 < \alpha < 1$.

Exercise 3.6 *Show that estimating equation (3.25) is unbiased for the AR(1), exchangeable, and tri-diagonal structures, but not for the unstructured structure.*

Exercise 3.7 *Verify that the inverse of the* $n_i \times n_i$ *AR(1) structure* $R_i(\alpha)$ *can be expressed as Equation (3.27). (Hint: This is most easily shown via direct multiplication.)*

Exercise 3.8 *Prove that the stage one QLS estimator for the AR(1) structure (3.31) will always take value in the positive-definite interval* $(-1, 1)$ *for the AR(1) structure.*

Exercise 3.9 *Prove that the stage two estimator* $\widehat{\alpha}_{QLS}$ *of* α *that is defined in Equation (3.37) will always take value in the positive-definite interval* $(-1, 1)$ *for the AR(1) structure.*

Exercise 3.10 *Assume that the working structure* $R_i(\alpha)$ *is AR(1). Show that the solution (for* α) *to Equation (3.38) will be the same for the AR(1), tri-diagonal, and exchangeable structures.*

Chapter 4

Mixed Linear Structures and Familial Data

Medical studies often involve measurements that are collected on unrelated subjects. However, some investigators are interested in relationships among family members, especially when studying genetic versus environmental factors. For example, Feng et al. (2009) proposed an approach for assessing family risk that incorporates the relationship among family members. Magnus et al. (2001) studied the similarity of birth-weight between family members (mother–child, father–child, and mother–father). These authors were especially interested in estimating the father–child correlations, which they reported as being less confounded by nongenetic effects than mother–child correlations.

In this chapter we describe the implementation of QLS for analysis of familial data from nuclear families that may vary in size and composition. We describe several familial correlation structures that are members of the class of mixed linear correlation structures. We then implement these structures using QLS and demonstrate the application of both QLS and GEE in an analysis of familial data using Stata software. We end with a small simulation study that demonstrates that QLS can estimate the regression parameter more efficiently than GEE in analysis of familial data.

Xie et al. (2010) implemented QLS for familial data. We present their results for the familial structure before describing earlier implementations of QLS for the basic structures described in Section 2.3.3 of Chapter 2. We describe QLS for the familial structures first because the more recent results of Xie et al. (2010) are applicable when describing the more basic structures. In particular, the results in Xie et al. (2010) make it possible to simplify some of the earlier derivations for the basic structures.

4.1 Notation for Data from Nuclear Families

In this chapter we consider data from nuclear families. We consider the situation in which we have one measurement per family member, so that y_{ij} now represents the measurement on member j of family i. Families may contain none, one, or both parents, and zero or more children. We will distinguish between the father, mother, and children (which we refer to as children or siblings) and will sort the vectors of measurements on each family so that the members are presented in the order of father, mother, child. For example, in Section 1.5.5 Magnus et al. (2001) considered families that contained one father, one mother, and one child. The vector of birth-weights on family i would be represented by (y_{i1}, y_{i2}, y_{i3}), where y_{i1} is the birth-weight of the father, y_{i2} is the birth-weight of the mother, and y_{i3} is the birth-weight of the child.

The study considered by Magnus et al. (2001) had the same composition for all families, because each family contained both parents and one child. However, in studies for which the composition varies, the measurements will be ordered within vectors so that the father is followed by the mother, who is followed by the children in no particular order.

4.2   Familial Correlation Structures for Analysis of Data from Nuclear Families

Earlier in this text we considered working correlation structures that were a function of a one-dimensional correlation parameter α. Now, when describing the association between members of nuclear families that may include parents and their off-spring, we will need to use additional parameters. We assume that the father–mother, father–child, mother–child, and child–child correlations are

- α_p (with subscript p for *parent*),
- α_f (with subscript f for *father*),
- α_m (with subscript m for *mother*),
- and α_c (with subscript c for *child*), respectively.

We then define the vector of parameters $\boldsymbol{\alpha} = (\alpha_p, \alpha_f, \alpha_m, \alpha_c)'$ and matrix W_i, which will be used to construct a vector of parameters $\boldsymbol{\alpha}_i$ for each family as follows. First, for family i, construct the matrix W_i by removing rows from a 4×4 identity matrix that correspond to the parameters that are *not involved* in family i. For example, if the ith family includes both parents and only one child, then the correlation structure for that family will not involve the parameter α_c that represents child–child correlation. The matrix W_i for family i will then be constructed by removing the fourth row of the 4×4 identity matrix. The vector $\boldsymbol{\alpha}_i$ of parameters for family i will then be defined as $\boldsymbol{\alpha}_i = W_i \boldsymbol{\alpha}$, or

$$\boldsymbol{\alpha}_i = \begin{pmatrix} 1 & 0 & 0 & 0 \\ 0 & 1 & 0 & 0 \\ 0 & 0 & 1 & 0 \end{pmatrix} \begin{pmatrix} \alpha_p \\ \alpha_f \\ \alpha_m \\ \alpha_c \end{pmatrix} = \begin{pmatrix} \alpha_p \\ \alpha_f \\ \alpha_m \end{pmatrix}. \tag{4.1}$$

Working correlation structures to describe the within family associations can then be constructed as follows.

If family i includes both parents and children, this results in an *extended familial correlation structure* $R_i(\boldsymbol{\alpha}_i)$ that can be used to describe the pattern of association among the n_i measurements on family i:

$$R_i(\boldsymbol{\alpha}_i) = \begin{pmatrix} 1 & \alpha_p & \alpha_f & \alpha_f & \cdots & \alpha_f \\ \alpha_p & 1 & \alpha_m & \alpha_m & \cdots & \alpha_m \\ \alpha_f & \alpha_m & 1 & \alpha_c & \cdots & \alpha_c \\ \alpha_f & \alpha_m & \alpha_c & 1 & \cdots & \alpha_c \\ \vdots & \vdots & \vdots & \vdots & \ddots & \vdots \\ \alpha_f & \alpha_m & \alpha_c & \alpha_c & \cdots & 1 \end{pmatrix}_{n_i \times n_i}, \tag{4.2}$$

where $\boldsymbol{\alpha}_i = W_i \boldsymbol{\alpha}$ for $W_i = I_4$, so that $\boldsymbol{\alpha}_i = I_4 \boldsymbol{\alpha} = \boldsymbol{\alpha}$ for each family with structure (4.2). The feasible region (assuming $n_i \geq 4$) for structure (4.2) can be expressed as (Chaganty and Deng, 2007, Equation 19)

$$-1 < \alpha_p < 1, \tag{4.3}$$
$$-1/(n_i - 3) < \alpha_c < 1,$$
$$(n_i - 2)(\alpha_f^2 + \alpha_m^2 - 2\alpha_p \alpha_f \alpha_m) < (1 - \alpha_p^2)(1 + (n_i - 3)\alpha_c).$$

Within the feasible region (4.3), the correlation structure (4.2) will always be positive definite. Therefore, when we refer to *feasible region* in this section, we refer to the values of the correlation parameters that yield a positive-definite correlation matrix. As noted earlier, additional constraints on the correlations may be required when the outcome variable is discrete.

Working structures for families with both parents and only one child, as given in Magnus et al. (2001), will not involve child–child correlations. The working structure for mother–father–child trios can be expressed as the following 3×3 correlation structure $R_i(\boldsymbol{\alpha}_i)$:

$$R_i(\boldsymbol{\alpha}_i) = \begin{pmatrix} 1 & \alpha_p & \alpha_f \\ \alpha_p & 1 & \alpha_m \\ \alpha_f & \alpha_m & 1 \end{pmatrix}, \tag{4.4}$$

where $\boldsymbol{\alpha}_i = (\alpha_p, \alpha_f, \alpha_m)'$ was defined in (4.1) for each family with structure (4.4). The feasible region for (4.4) can be expressed as

$$(\alpha_f, \alpha_p, \alpha_m) \in (-1, 1), \tag{4.5}$$
$$\alpha_f \alpha_p - h(\alpha_f, \alpha_p) < \alpha_m < \alpha_f \alpha_p + h(\alpha_f, \alpha_p), \tag{4.6}$$
$$\alpha_m \alpha_p - h(\alpha_m, \alpha_p) < \alpha_f < \alpha_m \alpha_p + h(\alpha_m, \alpha_p), \tag{4.7}$$
$$\alpha_f \alpha_m - h(\alpha_f, \alpha_m) < \alpha_p < \alpha_f \alpha_m + h(\alpha_f, \alpha_m), \tag{4.8}$$

where $h(\alpha_p, \alpha_f) = \sqrt{1 - \alpha_f^2 - \alpha_p^2 + \alpha_f^2 \alpha_p^2}$. Matrix (4.4) could also be obtained as the leading principal 3×3 sub-matrix of structure (4.2).

For families with only a mother and children, the working structure reduces to the *familial* structure $R_i(\boldsymbol{\alpha}_i)$, where

$$
R_i(\boldsymbol{\alpha}_i) = \begin{pmatrix}
1 & \alpha_m & \alpha_m & \cdots & \alpha_m \\
\alpha_m & 1 & \alpha_c & \cdots & \alpha_c \\
\alpha_m & \alpha_c & 1 & \cdots & \alpha_c \\
\vdots & \vdots & \vdots & \ddots & \vdots \\
\alpha_m & \alpha_c & \alpha_c & \cdots & 1
\end{pmatrix}_{n_i \times n_i},
\tag{4.9}
$$

where $\boldsymbol{\alpha}_i = W_i \boldsymbol{\alpha}$ for $W_i = \begin{pmatrix} 0 & 0 & 1 & 0 \\ 0 & 0 & 0 & 1 \end{pmatrix}$, so that $\boldsymbol{\alpha}_i = (\alpha_m, \alpha_c)'$ for each family with structure (4.9). Structure (4.9) could be obtained by deleting the first row and first column of (4.2) that has dimension $n_i + 1$.

Next, if a family only contains a father and children, then the working structure is *familial* $R_i(\boldsymbol{\alpha}_i)$, which can be defined in structure (4.9), but with α_m replaced with α_f in (4.9).

For families with only children, the correlation structure is *exchangeable* $R_i(\boldsymbol{\alpha}_i)$, where

$$
R_i(\boldsymbol{\alpha}_i) = \begin{pmatrix}
1 & \alpha_c & \cdots & \alpha_c \\
\alpha_c & 1 & \cdots & \alpha_c \\
\vdots & \vdots & \ddots & \vdots \\
\alpha_c & \alpha_c & \cdots & 1
\end{pmatrix}_{n_i \times n_i},
\tag{4.10}
$$

where $\boldsymbol{\alpha}_i = W_i \boldsymbol{\alpha}$ for $W_i = \begin{pmatrix} 0 & 0 & 0 & 1 \end{pmatrix}$, so that $\boldsymbol{\alpha}_i = \alpha_c$ for each family i with structure (4.10). The feasible region for this structure is

$$
-\frac{1}{n_i - 1} < \alpha_c < 1.
$$

Structure (4.10) could be obtained by deleting the first and second rows and first and second columns of (4.2) that has dimension $n_i + 2$.

Or, for families with both parents and no children, the correlation structure is the 2×2 *exchangeable* structure $R_i(\boldsymbol{\alpha}_i)$, where

$$
R_i(\boldsymbol{\alpha}_i) = \begin{pmatrix} 1 & \alpha_p \\ \alpha_p & 1 \end{pmatrix}_{2 \times 2},
\tag{4.11}
$$

where $\boldsymbol{\alpha}_i = W_i \boldsymbol{\alpha}$ for $W_i = \begin{pmatrix} 1 & 0 & 0 & 0 \end{pmatrix}$, so that $\boldsymbol{\alpha}_i = \alpha_p$ for each family i with structure (4.11). The feasible region for this structure is

$$
-1 < \alpha_p < 1.
$$

Structure (4.11) could also be obtained as the leading principal 2×2 sub-matrix of (4.2).

In practice, we might anticipate that structure (4.11) is a 2×2 identity structure, or more generally that $\alpha_p = 0$, because measurements on two parents are typically not related due to genetic similarity. (In other words, hopefully the parents are not close relatives.) However, we might anticipate similarity between two parents due to environmental reasons. For example, a couple who lives together might have similar eating patterns. If they overeat (or undereat), they might both tend to be heavy (or lean), so that their weights will be positively correlated. We might anticipate negative association when food is limited. For example, if two rats in a cage are deliberately underfed, the male rat who overeats will do so at the expense of the female rat, so that as he gets fatter, she becomes more lean.

4.3 Other Work on Assessment of Familial Correlations with QLS

Deng, Sabo, and Chaganty have been doing some important work on familial data. Sabo and Chaganty (2009) implemented QLS for structure (4.2) for continuous outcomes and for families of equal size and composition. They compared the QLS estimators of the correlation parameters with maximum likelihood and moment estimators for multivariate normal data. They showed that the QLS estimators were nearly as efficient as the maximum likelihood estimators in the large sample case for Gaussian data. For finite samples, they used simulations for Gaussian data to demonstrate that the quasi-least squares estimators can be more robust than the moment (GEE) or maximum likelihood estimators with respect to the estimated mean square error and infeasibility probabilities (of obtaining non-positive definite estimated correlation matrices).

Sabo and Chaganty (2010a) then implemented an autoregressive correlation structure in the analysis of balanced familial clustered data, for families with one parent and the same number of children per family. For continuous outcomes, they showed that the quasi-least squares estimators are nearly as efficient as the maximum likelihood estimators. For the small sample case, they again simulated multivariate normal data and demonstrated that the quasi-least squares estimators can be more robust than the maximum likelihood estimators in estimation of the correlation parameters.

Sabo and Chaganty (2010b) compared several hypothesis testing procedures for the parameters in a one parent autoregressive structure and in the nuclear family (4.2), for continuous outcome variables. They used likelihood ratio and Wald's tests for maximum likelihood estimators, and Wald-type tests for quasi-least squares and GEE estimators. They compared the different testing procedures via simulations for data from a multivariate normal distribution. Their tests could be helpful in assessing intra-familial associations. For example, they considered a test of the hypothesis $\alpha_p = 0$ to determine if the correlation between parents is significantly different from zero.

Chaganty and Deng (2007) considered binary outcomes with familial patterns of association. For binary variables, there are additional constraints for measures of association that are necessary to ensure the existence of a valid multivariate distribution for the data; Chaganty and Deng (2007) obtained the constraints for some three-dimensional marginal binary distributions and several measures of association, including kappas, odds ratios, relative risks, and correlations.

4.4   Justification of Implementation of QLS for Familial Structures via Consideration of the Class of Mixed Linear Correlation Structures

Xie et al. (2010) justified implementation of QLS for familial structures by proving general results for a larger class of correlation structures, the class of *mixed linear correlation structures*.

4.4.1   Definition of Mixed Linear Correlation Structures

Mixed correlation structures are structures that can vary between families. *Linear* correlation structures are structures whose off-diagonal elements are linear functions of the correlation parameters. For example, suppose we consider a study that included twenty mother–father–child trios, and thirty families with both parents and no children. Plausible correlation structures for this study might be represented by structure (4.4) for the trios, and structure (4.11) for the families without children. These structures are *mixed* because they vary in composition between families; they are *linear* because their off-diagonal elements are linear functions of α_p, α_f, α_m, or α_c. Other examples of linear correlation structures include all the familial structures defined in Section 4.2. In addition, the exchangeable and tri-diagonal structures defined in Section 2.3.3 are linear structures. Another linear structure is the banded Toeplitz structure that was implemented in Shults et al. (2006a). An example of a correlation structure that *is not* linear is the AR(1) structure defined in Section 2.3.3; element (j,k) of this structure is $\alpha^{|j-k|}$, which clearly is not a linear function of α.

Next we give a formal definition of a *linear correlation structure*. Suppose that α_i is an $s_i \times 1$ vector. We refer to $n_i \times n_i$ correlation matrix $R_i(\alpha_i)$ as **linear** if

$$R_i(\alpha_i) = \sum_{j=1}^{s_i} \left(R_i(o_j) - I_{n_i} \right) \alpha_i[j] + I_{n_i}, \tag{4.12}$$

where o_j is a unit vector with only the jth entry equal to 1, and all other entries equal to 0; $\alpha_i[j]$ is the jth element of α_i; and I_{n_i} is an $n_i \times n_i$ identity matrix. For example, structure (4.11) is a linear correlation structure because it can be expressed as

$$R_i(\alpha_i) = (R_i(o_1) - I_2)\,\alpha_i[1] + I_2 \tag{4.13}$$

$$= \left(\begin{pmatrix} 1 & 1 \\ 1 & 1 \end{pmatrix} - \begin{pmatrix} 1 & 0 \\ 0 & 1 \end{pmatrix} \right) \alpha_p + \begin{pmatrix} 1 & 0 \\ 0 & 1 \end{pmatrix} \tag{4.14}$$

$$= \begin{pmatrix} 1 & \alpha_p \\ \alpha_p & 1 \end{pmatrix}. \tag{4.15}$$

Note that vector α_i is identifiable if and only if the following condition holds:

$$\sum_{j=1}^{s_i} (R_i(o_j) - I_{n_i})c_j = 0 \text{ if and only if } c = (c_1, \dots, c_{s_i})' = 0, \tag{4.16}$$

where $c = (c_1, \dots, c_{s_i})'$ is a vector of constants.

We define the feasible region Ω to be the intersection of the feasible regions Ω_i for all structures, so that $\Omega = \cap_{i=1}^{m} \Omega_i$ is the interval on which all structures are positive definite.

4.4.2 Results for General Correlation Structures (for Stage One of QLS) and for Linear Correlation Structures (for Stage Two of QLS)

The results in this section are general, and can be applied to more than just familial data. For this reason, rather than refer to families, here we refer to clusters of measurements. A cluster could represent a family, or it could something else, for example, a set of repeated measurements on one subject.

4.4.2.1 Results for Stage One

Xie et al. (2010) developed general results that removed the need to prove feasibility of the QLS stage one estimators on a case-by-case basis, as was done previously. We state their results here. See Xie et al. (2010) for the proofs of the results presented in this chapter.

First, recall that QLS estimates the correlation parameters in two stages. In the first stage, QLS obtains estimates by minimizing an objective function, the generalized error sum of squares, that was defined in Equation (3.13). The next theorem establishes that the global point of minimum of the generalized error sum of squares will always take a value in the feasible region Ω.

Theorem 4.1 *[From Xie et al. (2010)] If for each subject i, $R_i(\boldsymbol{\alpha}_i)$ is a differentiable $n_i \times n_i$ matrix, then the global minimizer of the generalized error sum of squares $Q^*(\alpha, \beta)$ that was defined in Equation (3.13) is an inner point of Ω, where Ω is the feasible region of $(R_i(\boldsymbol{\alpha}_i))_{1,\ldots,m}$.*

Theorem 4.1 establishes that the global minimum of the generalized error sum of squares will yield estimated correlation matrices that are positive definite, which is important to lessen the chance of a convergence failure in the estimation procedure.

The global minimum of the generalized error sum of squares was obtained by taking the first derivative with respect to the correlation parameter and equating with zero, to obtain the stage one estimating equation (3.40). However, while Theorem 4.1 ensures the existence of solutions for Equation (3.40), it does not guarantee that the roots of this equation are unique, in which case it may be difficult to obtain all roots, or to search among several local points of minimum to obtain the global minimum.

However, the next theorem establishes that under a fairly general condition, the generalized error sum of squares will be convex almost surely, so that there will exist a unique root for the stage one estimating equation (3.40) almost surely.

Theorem 4.2 *[From Xie et al. (2010)] Suppose each cluster $i \in \{1,\ldots,m\}$ in the data under consideration has correlation structure $R_i(\boldsymbol{\alpha}_i)$. If $\forall \ \boldsymbol{\alpha}_i \in \Omega$,*

$$\sum_{j=1}^{s_i} \frac{\partial R_i(\boldsymbol{\alpha}_i)}{\partial \alpha_j} c_j = 0 \text{ if and only if } \boldsymbol{c} = (c_1,\ldots,c_{s_i})' = 0, \tag{4.17}$$

then Equation (3.40) has a unique solution in the feasible region Ω almost surely.

Linear correlation structures satisfy the condition (4.17). As a consequence, the following result follows easily.

Corollary 4.3 *[From Xie et al. (2010)] Suppose for each cluster or subject $i \in \{1,\ldots,m\}$ we have a linear correlation structure $R_i(\boldsymbol{\alpha}_i)$ of the form (4.12). Then if $\boldsymbol{\alpha}_i$ is identifiable, the QLS stage one estimating equation (3.40) has a unique solution in the feasible region Ω almost surely.*

The results provided in Xie et al. (2010) for stage one of QLS establish that the earlier proofs that were provided to prove the *existence* of the stage one estimators on a case-by-case basis were not necessary. In addition, the criterion (4.17) provided in Theorem 4.2 that ensures the *uniqueness* of the stage one estimators is fairly general and is satisfied by several structures, including the exchangeable, tri-diagonal, and unstructured patterns that were defined in Chapter 2.3.3; the banded Toeplitz structure defined in Shults et al. (2006a); and the familial structures that we consider in this chapter.

4.4.2.2 Results for Stage Two

Recall that although the stage one QLS estimator is feasible, it is not consistent. A consistent estimator is obtained in stage two of the procedure, by solving the stage two estimating equation (3.41). For the stage two estimators, the results of Xie et al. (2010) are not as general. The following theorem establishes that for mixed *linear* correlation structures, the stage two estimator exists and is unique, with probability one.

Theorem 4.4 *[From Xie et al. (2010)] If for each cluster $i \in \{1,\ldots,m\}$, the within subject correlation $R_i(\boldsymbol{\alpha}_i)$ has a linear correlation structure of form (4.12), then the stage two estimating equation (3.41) has a unique solution with probability one.*

Theorem 3.2 of Chaganty and Shults (1999) establishes that if there is a unique root to the stage two estimating equation (3.41) that is a continuous and one-to-one function of the stage one estimator, then this solution will be a consistent estimate of the correlation parameter. Theorem 4.4 establishes that the stage two estimators will be consistent for linear correlation structures, which, as mentioned above, include several structures in addition to the familial structures that were defined in Section 4.2.

However, note there is an important distinction between the results for the stage one and stage two estimators. They guarantee that stage one of QLS will yield *feasible estimates* for all working structures, while stage two will yield *consistent* estimators for the class of mixed linear correlation structures. This means that only stage one guarantees an estimated correlation matrix that is positive definite for any sample size, unless we provide an additional proof for a particular structure that the stage two estimate will always be feasible (as we did for the AR(1) structure). This is an important distinction that we will revisit when we discuss the tri-diagonal structure in Chapter 5.

In the proof of Theorem 4.4, Xie et al. (2010) provided an approach that can be used to obtain an explicit solution to the QLS stage two estimating equation for linear correlation structures. Although the restriction is not necessary, it is most straightforward to express the results in a simple form if we assume that $\alpha_i = \alpha \; \forall \; i$. Suppose we obtain the stage one estimator $\widehat{\alpha}_{QONE}$. Define $A_{ij} = R_i^{-1}(\widehat{\alpha}_{QONE})(R_i(o_j) - I_{n_i}))$, $M_{jk} = \sum_{i=1}^{m} \text{tr}(A_{ij}A_{ik})$, and $w_j = -\sum_{i=1}^{m} \text{tr}(A_{ij}R_i^{-1}(\widehat{\alpha}_{QONE}))$. Let $M(\widehat{\alpha}_{QONE}) = (M_{jk})_{s \times s}$ and $w(\widehat{\alpha}_{QONE}) = (w_1, \ldots, w_s)'$. We can then express the stage two estimator in a very simple form:

$$\widehat{\alpha}_i = M^{-1}(\widehat{\alpha}_{QONE})w(\widehat{\alpha}_{QONE}). \tag{4.18}$$

4.5 Demonstration of QLS for Analysis of Balanced Familial Data Using Stata Software

Here we consider familial data described in Section 1.5.5 of Chapter 1 that includes birth-weights (in grams) and related information on 1,000 mother–father–child trios. We demonstrate implementation of QLS with the working structure for (mother, father, child) trios that was defined in structure (4.4). Structure (4.4) is identical to a 3×3 unstructured correlation matrix, so that we are also demonstrating implementation of a three-dimensional unstructured matrix.

We use the Stata user authored command **xtqlsfam6.ado** (Shults) to implement structure (4.4) in a regression of birth-weight on male gender (*male*), first born status (*first*), number of years since 1967 (*byr*), and whether the child's mother was between 20 and 35 (*midage*), or greater than 35 (*highage*). For comparison, we also implement GEE with Stata's built-in **xtgee** command, for both unstructured and exchangeable structures. All analysis results in this section were obtained using a sandwich estimate of the covariance matrix.

First, we fit QLS with working structure (4.4).

```
. xtqlsfam6  bwt male first byr midage highage, /*
      */ i(family) t(type) c(trio) f(gau) vce(robust)

Iteration 1: tolerance = .31949001
Iteration 2: tolerance = 1.267e-13
```

GEE population-averaged model			Number of obs	=	3000
Group and time vars:	family __00001C		Number of groups	=	1000
Link:		identity	Obs per group: min =		3
Family:		Gaussian	avg =		3.0
Correlation:	fixed (specified)		max =		3
			Wald chi2(5)	=	161.53
Scale parameter:		233090.7	Prob > chi2	=	0.0000

(Std. Err. adjusted for clustering on family)

bwt	Coef.	Semirobust Std. Err.	z	P>|z|	[95% Conf. Interval]	
male	158.1303	17.91808	8.83	0.000	123.0115	193.2491
first	-139.1568	18.83117	-7.39	0.000	-176.0652	-102.2484
byr	3.639571	.6847018	5.32	0.000	2.29758	4.981562
midage	56.44828	30.63661	1.84	0.065	-3.598368	116.4949

```
   highage |   118.4659   59.92465     1.98   0.048    1.015732    235.916
     _cons |     3218.2   57.54763    55.92   0.000    3105.409   3330.992
--------------------------------------------------------------------------
```

The results suggest that being male, not first born, having a mother > 35 years of age, and being born more recently, are all significantly associated with having a higher birth weight. Next we examine the estimated working correlation structure.

. xtcorr

Estimated within-family correlation matrix R:

```
          c1        c2        c3
r1   1.0000
r2   0.0077    1.0000
r3   0.1966    0.2374    1.0000
```

The estimated correlation between the birth-weight of parents is negligible (0.0077) while the correlation between father and child is slightly smaller than the correlation between mother and child. This is compatible with the hypothesis that similarity in birth-weights is determined by genetic relationship, so that birth-weights are more similar between parent and child (and slightly more so for mothers versus fathers) than between spouses who are not genetically related.

Next we implement GEE with the same working structure (4.4) that is equivalent to a 3×3 unstructured matrix.

. xtgee bwt male first byr midage highage, i(family) t(type) c(uns) f(gau) robust

```
Iteration 1: tolerance = .32159109
Iteration 2: tolerance = .0044032
Iteration 3: tolerance = .000041
Iteration 4: tolerance = 3.841e-07
```

```
GEE population-averaged model           Number of obs      =       3000
Group and time vars:      family type   Number of groups   =       1000
Link:                        identity   Obs per group: min =          3
Family:                      Gaussian                  avg =        3.0
Correlation:             unstructured                  max =          3
                                        Wald chi2(5)       =     161.57
Scale parameter:             233093.4   Prob > chi2        =     0.0000
```

(Std. Err. adjusted for clustering on family)

bwt	Coef.	Semirobust Std. Err.	z	P>|z|	[95% Conf. Interval]	
male	158.1844	17.92141	8.83	0.000	123.059	193.3097
first	-139.3524	18.83839	-7.40	0.000	-176.2749	-102.4298
byr	3.639129	.6847313	5.31	0.000	2.297081	4.981178
midage	56.01637	30.63153	1.83	0.067	-4.020328	116.0531
highage	117.4442	59.9238	1.96	0.050	-.0042549	234.8927
_cons	3218.694	57.54555	55.93	0.000	3105.906	3331.481

```
. xtcorr

Estimated within-family correlation matrix R:

        c1      c2      c3
r1   1.0000
r2   0.0088  1.0000
r3   0.2042  0.2377  1.0000
```

The analysis results were almost identical for QLS versus GEE. However, in the next section we show that for larger correlations, QLS can estimate the regression parameter more efficiently than GEE.

Next we implement GEE with the exchangeable working structure; as noted earlier, the exchangeable structure is popular for analysis of familial data.

```
. xtgee bwt male first byr midage highage, i(family) t(type) c(exc) f(gau) robust

Iteration 1: tolerance = .228722
Iteration 2: tolerance = .00243847
Iteration 3: tolerance = .00001979
Iteration 4: tolerance = 1.602e-07

GEE population-averaged model          Number of obs      =       3000
Group variable:               family   Number of groups   =       1000
Link:                       identity   Obs per group: min =          3
Family:                     Gaussian                  avg =        3.0
Correlation:             exchangeable                  max =          3
                                        Wald chi2(5)       =     165.50
Scale parameter:               233063   Prob > chi2        =     0.0000

                      (Std. Err. adjusted for clustering on family)
------------------------------------------------------------------------
             |              Semirobust
        bwt  |     Coef.    Std. Err.      z    P>|z|   [95% Conf. Interval]
-------------+----------------------------------------------------------
       male  |  160.8722   17.88988     8.99   0.000    125.8087   195.9357
      first  |   -142.16   18.97276    -7.49   0.000   -179.346  -104.9741
        byr  |  3.661427   .6865331     5.33   0.000    2.315847   5.007008
     midage  |  68.05747   30.99067     2.20   0.028    7.316867   128.7981
    highage  |  127.8645   60.1534      2.13   0.034    9.966029    245.763
      _cons  |   3206.18   57.79844    55.47   0.000    3092.897   3319.463
------------------------------------------------------------------------

. xtcorr

Estimated within-family correlation matrix R:

        c1      c2      c3
r1   1.0000
r2   0.1501  1.0000
r3   0.1501  0.1501  1.0000
```

As might be anticipated from the estimated correlations for the familial structure (which were small), the estimated correlations are also small for the exchangeable structure. However, it is interesting to note that having a mother between 20 and 35 years of age is only a significant correlate of higher birth-weight for the exchangeable structure (because the p-value for *midage* is < 0.05 for this structure), but not for the familial structure. In their analysis of these data with a mixed model, Magnus et al. (2001) also failed to identify *midage* as a significant correlate of higher birth-weight. Because the familial and exchangeable structures gave different results, it will be interesting to compare the fit of these structures in Chapter 8 when we discuss methods for comparing the fit of several working structures.

4.6 Demonstration of QLS for Analysis of Unbalanced Familial Data Using R Software

In Section 4.5 we presented an analysis of a balanced familial dataset comprising 1,000 mother–father–child trios. Currently, the only software available for analysis of unbalanced familial data is the R package QLSPACK (Xie and Shults, 2009). Here we demonstrate application of this software in an analysis of a sub-sample of that data presented in Section 4.5, that was obtained by randomly dropping some measurements.

Here is a summary of the output of that analysis.

```
> summary(qls.obj)
```

```
Call: mixed.exch.fam.exfam1(formula = bwt ~ male + first +
mothage + midage + highage, data = dat, id1 = id1, id2 = id2,
id3 = id3, id4 = id4, family = gaussian, time = NULL,
std.err = "san.se")


Coefficients:  Estimate  Std.err  Wald     Pr(>|W|)
(Intercept)    3361.623  76.177   1947.370 < 2e-16  ***
male            138.782  19.429     51.023 9.13e-13 ***
first           -91.414  22.054     17.181 3.40e-05 ***
mothage           6.060   3.154      3.693 0.0546
midage           69.623  39.302      3.138 0.0765
highage          48.437  89.008      0.296 0.5863
---
Signif. codes: 0 *** 0.001 ** 0.01 * 0.05 . 0.1   1

Estimated Scale Parameters: Estimate Std.err (Intercept)
                    238506    8598


Correlation: Structure = exch fam exfam Link = identity

Estimated Correlation Parameters:  Estimate Std.err
                        alpha:1  0.05949 0
                        alpha:2 0.12446 0
                        alpha:3 0.17112 0

Number of clusters: 999 Maximum cluster size: 6
```

As can be seen from the output above, there are 999 clusters (families) in the analysis, so that randomly dropping measurements from the balanced dataset resulted in the loss of one complete family. The maximum family size was six. The estimated regression coefficients were similar to those for the complete dataset.

4.7 Simulations to Compare Implementation of QLS with Correct Specification of the Trio Structure versus Correct Specification with GEE and Incorrect Specification of the Exchangeable Working Structure with GEE

In this section we conduct a simulation study in Stata to assess the implications of fitting an exchangeable correlation structure with GEE, when the true correlation structure is familial (4.4). Because the structure (4.4) is a 3×3 unstructured matrix, we also implemented a 3×3 unstructured matrix using the moment estimates that are implemented in the **xtgee** command in Stata. The number of measurements per subject was small and was constant for all subjects, so that conditions were ideal for application of an unstructured matrix, which coincides with the familial structure (4.4) when all families consist of (mother, father, child) trios.

We simulated data for the model that was implemented for analysis in Section 4.5, with assumed parameter values that were very close to those obtained for the QLS analysis with the familial structure. We assumed the following values for the regression coefficients for each variable: *158.1303* for *male*; -139.1568 for *first*; 3.639571 for *byr*; 56.44828 for *midage*; 118.4659 for *highage*; and 3218.2 for *cons*. We also assumed the following values for the correlation parameters: $(\alpha_p, \alpha_f, \alpha_m)' = (0.0077, 0.1966, 0.2374)$. Furthermore, in addition to the previous values that were very close to the estimates obtained in Section 4.5, we also considered larger values for the (father,child) and (mother,child) correlations: $(\alpha_p, \alpha_f, \alpha_m)' = (0, 0.64, 0.75)$. Setting the (father,mother) correlations equal to zero in the latter values is biologically plausible because there is usually no close genetic relationship between parents. In addition, specifying zero correlation between parents and relatively large (father,child) and (mother,child) correlations results in a correlation structure that differs from an exchangeable structure to a greater degree than did the structure with $(\alpha_p, \alpha_f, \alpha_m)' = (0.0077, 0.1966, 0.2374)'$.

The dataset we considered in the previous section included 1,000 (father,mother,child) trios and thus was quite large. We retained the values of the covariates for the simulations, but simulated correlated errors during each simulation run according to the following algorithm.

Algorithm for the simulations:

1. Open up the familial dataset that was described in Section 1.5.5 of Chapter 1. Sort on family identification variable and the variable *type* that indicates order within the families. After sorting, the fathers will be listed first within each family, followed by mother, and then child.

2. Next, simulate a new variable that includes independent normally distributed observations with mean zero and variance equal to some specified value; we chose $\phi = 233,090.7$, which was the estimate obtained in the GEE analysis of these data.

3. Next, to obtain transformed errors with a true familial structure, multiply the errors obtained in Step 2 by the square root of familial structure (4.4) that is evaluated at the assumed values of the correlations. To speed up the computations, we made use of the fact that if $R^{1/2}[j,k]$ is element (j,k) of the square root of the working structure and $(e_{i1}, e_{i2}, e_{i3})'$ is the vector of errors on family i, then the transformed values can be expressed as $(e_{i1}^*, e_{i2}^*, e_{i3}^*)'$, where $e_{ia}^* = R^{1/2}[a,1]e_{i1} + R^{1/2}[a,2]e_{i2} + R^{1/2}[a,3]e_{i3}$ for $a \in \{1,2,3\}$.

4. Next, to obtain the outcome variable for the simulations, add $\mu_{ij} = x_{ij}'\beta$ to the transformed errors obtained in Step 3, to obtain the simulated outcomes $y_{ij} = \mu_{ij} + e_{ij}^*$.

5. Next, implement QLS and GEE in a regression of the simulated outcome variable on the covariates involved in x_{ij}. The working correlation structures in the GEE and QLS regressions include the 3×3 familial structure (which is the same as an unstructured matrix in this case), and GEE with an exchangeable structure, in a regression of the simulated outcomes on the covariates *male*, *first*, *byr*, *midage*, and *highage*. (We refer to these approaches as QLS-UNS, GEE-UNS, and GEE-EXC, respectively.) Save the estimated regression coefficients and correlation parameters for each approach.

6. Repeat the previous Steps 1 through 5 a large number of times. (We used $2,000$ simulation runs.)

7. After running the simulations, compare the estimated regression coefficients for each approach. We compared the approaches with respect to bias, mean square error (MSE), and small sample efficiency. The bias for estimator $\hat{\theta}$ is defined as $\frac{1}{R}\sum_{i=1}^{R}(\theta - \hat{\theta}_i)$. The MSE for estimator $\hat{\theta}$ is defined as $\frac{1}{R}\sum_{i=1}^{R}(\theta - \hat{\theta}_i)^2$, where $R = 2,000$ is the number of simulation runs and θ is the true value. The small sample efficiency for each approach is defined as the ratio of the MSE for each approach to the MSE for QLS. An efficiency greater than 1 indicates that the MSE for QLS is smaller than that of the method under consideration, so that QLS is superior with respect to MSE in this situation.

Table 4.1 displays the small sample efficiencies for smaller and larger values of the correlation parameters. Rows one and three of Table 4.1 display efficiencies for GEE with an incorrectly specified exchangeable (GEE-EXC) relative to QLS with a correctly specified familial structure; they allow for comparison of GEE and QLS when the true structure is correctly specified as familial for QLS, but is incorrectly specified as exchangeable for GEE.

Rows two and four of Table 4.1 display efficiencies for GEE with an unstructured working structure (GEE-UNS) relative to QLS with a correctly specified familial structure. Because the unstructured and familial structures are identical for this simulation scenario, rows two and four allow for comparison of QLS versus GEE under correct specification of the familial structure.

First we consider smaller values of the correlation. The values in row two of Table 4.1 were 1, which indicates that GEE and QLS were equally efficient when the true familial structure was correctly specified. The values in row one were only very slightly greater than 1, which indicates that correct specification of the true familial

Table 4.1 *Small Sample Efficiencies for Regression Parameters for Smaller and Larger Values of the Intra-Familial Correlations.*

Correlation	Method	*first*	*male*	*byr*	*midage*	*highage*
small	GEE-EXC	1.01	1.02	1.01	1.02	1.02
	GEE-UNS	1.00	1.00	1.00	1.00	1.00
large	GEE-EXC	2.09	1.38	1.16	2.16	2.20
	GEE-UNS	1.16	1.12	1.02	1.14	1.22

structure only yielded a very small gain in efficiency. However, when the correlations are small, the true familial and working exchangeable structures are close to an identity matrix and are therefore close in value; it is perhaps not surprising that correct specification of the familial structure only yielded a very small gain in efficiency because the incorrectly specified structure was close in value to the true structure in this situation.

Next we consider larger values of the correlation parameters. The values in row four of Table 4.1 were slightly greater than 1, which indicates that QLS was more efficient than GEE when the true familial structure was correctly specified for both QLS and GEE. For example, the efficiency for *highage* in row four was 1.22, which indicates that the MSE for GEE was 1.22 times the MSE for QLS for this variable. The efficiencies ranged from 1.02 to 1.22 when the true structure was correctly specified for both QLS and GEE. Next, the values in row three ranged from 1.16 to 2.20, which indicates that the MSE for GEE was between 1.16 and 2.20 times greater for GEE relative to QLS when the true familial structure was misspecified as exchangeable for GEE but was correctly specified for QLS.

The losses in efficiency were perhaps greater for row three versus row one of Table 4.1 because the difference between the true familial and working exchangeable structure was greater for higher values of the correlation; we were therefore making a bigger "mistake" in incorrectly applying an exchangeable structure when the values of the correlation were greater.

In addition, the bias for the regression parameters was relatively small and similar for all approaches. We also observed that no methods failed to converge and that for all methods, the estimated correlations took value in the feasible region defined in (4.5). All methods probably performed quite well with respect to convergence and feasibility because the sample size was relatively large for the simulations considered in this section.

4.8   Summary and Future Research Directions

In this chapter we considered familial data structures that are members of the class of mixed linear correlation structures. Xie et al. (2010) proved that a stage one QLS estimator will always exist for any structure, and furthermore, that a unique stage two estimator will always exist for a mixed linear correlation structure. The class of mixed linear correlation structures includes the familial structures considered in this chapter, in addition to other structures, including the unstructured, exchangeable, tri-diagonal, and banded Toeplitz.

For those who might be interested in further research in this area, some potential problems might include the following. We considered nuclear families but it could be of interest (and could be relatively straightforward) to extend the familial structures to incorporate information on additional generations of family members, and to develop software for implementation of those structures.

Current implementations of QLS in existing software do not allow the correlations to depend on values of the covariates, aside from the Markov structure to be discussed in Chapter 5 that directly involves the timings of measurements in the correlation structure. It could be of interest to allow the correlations to depend on the covariates in a more general way. For example, Kumanyika et al. (2009) compared weight loss between interventions for weight loss that employed varying degrees of social support. One stratum of Kumanyika's study randomized clusters of subjects to either a high or low social support group. If high versus low social support was more effective, we might anticipate greater weight loss in both the index and co-participants in the high social support group, which would be reflected in greater index–co-participant correlation of weight loss in that group. Shults and Morrow allowed the correlations to vary between groups in Shults and Morrow (2002). An interesting example using the birth-weight data would be to see if the correlation between birth-weights was stronger for younger parents. However, this generalization and further direct reliance of the correlations on covariates has not been made available in the software that is currently available for QLS.

4.9 Exercises

Exercise 4.1 *Consider a 2×2 correlation structure with parameter α. Show that this structure is a linear correlation structure, according to Equation (4.12).*

Exercise 4.2 *Prove that*

$$\frac{\partial R^{-1}(\alpha)}{\partial \alpha} = -R^{-1}(\alpha) \frac{\partial R(\alpha)}{\partial \alpha} R^{-1}(\alpha). \tag{4.19}$$

Exercise 4.3 *Consider a 4×4 exchangeable correlation structure that was defined in (2.11). Prove that Equation (4.18) does indeed provide the solution to the stage two estimating equation (3.41) for this structure.*

Exercise 4.4 *Consider a 3×3 unstructured correlation matrix that was defined in (2.14). Show that this structure is a linear correlation structure, according to Equation (4.12).*

Exercise 4.5 *Consider a 3×3 unstructured correlation matrix that was defined in Equation (2.14). Prove that the Equation (4.18) does indeed provide the solution to the stage two estimating equation (3.41) for this structure.*

Exercise 4.6 *Suppose we consider families that include a mother, father, two children, and a grandmother. Describe a familial correlation structure that might be plausible to describe the pattern of association among weights that are measured on each member of this family, at one measurement occasion.*

Chapter 5

Correlation Structures for Clustered and Longitudinal Data

In Chapter 3 we described the development of QLS that was originally presented in three manuscripts (Chaganty, 1997; Chaganty and Shults, 1999; Shults and Chaganty, 1998). Here we describe the implementation of QLS for the basic correlation structures first described in the seminal paper on GEE (Liang and Zeger, 1986), as well as for the Markov structure, which was not presented in Liang and Zeger (1986). The Markov correlation structure is appropriate for analysis of longitudinal data that are unequally spaced in time because it models the correlation as a function of the time gaps between measurements. In Section 5.1 we discuss the different ways that unequal spacing of data can arise.

We presented the algorithm for the two-stage QLS estimation procedure in Section 3.4 of Chapter 3. To briefly review, after selecting a starting estimate $\widehat{\alpha}_{QONE}$ for α, (e.g., $\widehat{\alpha}_{QONE} = 0$), stage one of QLS alternates between (1) updating $\widehat{\beta}$ by solving the GEE estimating equation (3.39) for β and (2) updating $\widehat{\alpha}_{QONE}$ by solving the stage one estimating equation (3.40) for α. After convergence in stage one, stage two obtains the final estimate $\widehat{\alpha}_{QLS}$ of α by solving the stage two estimating equation (3.41) for α. The final QLS estimate $\widehat{\beta}$ of β is then obtained by solving the GEE estimating equation (3.39) (evaluated at $\widehat{\alpha}_{QLS}$) for β.

In this chapter we provide details regarding estimation of α within each stage of the QLS estimation procedure, by describing how to solve the stage one and stage two estimating equations for each correlation structure that we consider. We provide some simplifications based on the results presented in Chapter 4, for structures that are linear correlation structures; however, in Chapter 4 the working structure for cluster i depended on an $s_i \times 1$ vector of correlation parameters $\boldsymbol{\alpha}_i$, while the structures in this chapter depend on the one-dimensional parameter α. Therefore, when we refer to results in Chapter 4, in the notation of that chapter, we assume that $\boldsymbol{\alpha}_i = \alpha$ and $s_i = 1$ for all i.

We then provide an algorithm for QLS estimation that can be used when software for GEE is available that allows for application of a user-specified correlation structure; this algorithm was used in the current implementations of QLS that are available in Stata (Shults et al., 2007) and R (Xie and Shults, 2009). In addition, we discuss what to do when software for GEE is not available or is not utilized, as was the case when software for QLS was developed in MATLAB (Ratcliffe and Shults, 2008) and in SAS (Kim and Shults, 2010), respectively.

5.1   Characteristics of Clustered and Longitudinal Data

In this chapter we discuss correlation structures that are appropriate for data that are equally or unequally spaced in time. We will first discuss the exchangeable structure that assumes all pairwise associations within subjects or clusters are equal. The exchangeable structure is often applied in cross-sectional studies, when all measurements are collected at approximately the same time. For example, Section 1.5.5 described a study that considered birth-weights within families. In this study, the birth-weights were all measured at approximately the same time (within a few minutes after birth). If we assumed that the correlations are constant between family members (e.g., between mother-father or mother-infant), then an exchangeable structure would be a plausible pattern of association for this study.

However, if the measurements are not collected at approximately the same time (e.g., at the start of a study or soon after birth), then it might be important to take the spacing of measurements into account. It is important to recognize that unequal spacing of measurements can occur in several ways, and that particular structures might be appropriate for particular patterns in the spacings. For example, different types of study designs can result in different patterns of timings of measurements:

1. **When the study design planned for the collection of equally spaced measurements at fixed time points, and the data are complete or some visits are missing for some subjects:** Many medical studies employ a design that involves measurement occasions at particular time points that are equally spaced. For example, in a clinical trial to compare two diets, the study design might call for measurement of body-weight at baseline and then at 6, 12, and 18 months post baseline. If all patients completed all their visits, then the spacing between any consecutive visits on a subject (e.g., between 12 and 18 months) would be 6 months.

   However, in most clinical trials, some subjects will miss some visits. If subjects drop out of the study and never return, for example, because they moved out of the study area, then the spacing between measurements will be retained, although the total number of measurements will vary between subjects. However, if some subjects miss a visit and then return, e.g., because they forgot about one particular visit, then the spacing between measurements will not be equal. For example, if some subjects are missing only the 6 month visit, then the data collection times on these subjects will be 0, 12, and 18 months, so that the spacing between the first and second observed measurements on these subjects will be 12 months. In studies with measurements taken at particular equally spaced time points, the timing of measurements as well as the time-gaps between observed measurements on a subject will be a subset of a set of possible times, or gaps between times, respectively.

2. **When the study design did not call for equal spacing of measurements but the planned measurement times are the same for all subjects:** Some study

designs will involve data collection at measurement times that are the same for all subjects, but that are unequally spaced. For example, a weight loss study might plan for body-weights to be measured at baseline, and then at 12 and 18 months post baseline. Other examples include the obesity data in Section 1.5.1 and hearing recognition data in Section 1.5.2. For subjects with complete data, consecutive measurements will not be equally spaced, but the time gaps will be the same for all subjects. For example, the time interval between the first and second measurements will be 6 months, among subjects with complete data.

3. **When the study design did not call for measurement times that are equally spaced or the same for all subjects:** Some study designs involve measurement times that are not planned to be the same across subjects. For example, in Preisser et al. (2003a) the researchers collected repeated information on migrant workers during an observation period of 83 consecutive work days. However, not all subjects worked every day, which resulted in measurements that were unequally spaced in time, with timings and time gaps between measurements that were not the same for all subjects.

Even for studies with complete data that planned for the same number of equally spaced visits on each subject, the timings may be unequally spaced, with time gaps that are not the same for all subjects. For example, consider a study that planned for measurements to be collected on subjects at baseline and then at 6 and 12 months post baseline. If the actual date of measurement is available, the time intervals between consecutive measurements on subjects can be calculated exactly. In many medical studies, the data of the 6-month visit will not occur at exactly 6 months post baseline, for example, but will occur within a few days (or weeks) of six months. Therefore, even in studies that planned for the collection of measurements that are equally spaced in time, the measurements may be only approximately spaced if the actual date of measurement is taken into account.

5.2   The Exchangeable Correlation Structure for Clustered Data

5.2.1   Solutions to the QLS Stage One and Stage Two Estimating Equations for α

The exchangeable structure was described in Section 2.3.3 of Chapter 2, and was defined in Equation (2.11). This structure assumes that all pairwise correlations within clusters are constant and is therefore plausible for cross-sectional studies with measurements that are correlated within clusters, for example, within litters of rats or within intact social groups such as schools, churches, workplaces, or communities. It might also be applied in longitudinal studies if no temporal decay in the correlation of measurements is anticipated. For example, if repeated weights are collected on a subject on each of 3 consecutive days, then an exchangeable structure would be appropriate if we anticipate that the correlation between weights measured on days 1 and 3 is the same as the correlation between weights measured on days 1 and 2.

The $n_i \times n_i$ exchangeable structure (2.11) is a linear correlation structure because it can be expressed in the form (4.12) for a linear correlation structure:

$$R_i(\alpha) = (R_i(\boldsymbol{o}_1) - I_{n_i})\alpha + I_{n_i}$$

$$= \left\{ \begin{pmatrix} 1 & 1 & 1 & \cdots & 1 \\ 1 & 1 & 1 & \cdots & 1 \\ \vdots & \vdots & \vdots & \ddots & \vdots \\ 1 & 1 & 1 & \cdots & 1 \end{pmatrix}_{n_i \times n_i} - \begin{pmatrix} 1 & 0 & 0 & \cdots & 0 \\ 0 & 1 & 0 & \cdots & 0 \\ \vdots & \vdots & \vdots & \ddots & \vdots \\ 0 & 0 & 0 & \cdots & 1 \end{pmatrix}_{n_i \times n_i} \right\} \alpha$$

$$+ \begin{pmatrix} 1 & 0 & 0 & \cdots & 0 \\ 0 & 1 & 0 & \cdots & 0 \\ \vdots & \vdots & \vdots & \ddots & \vdots \\ 0 & 0 & 0 & \cdots & 1 \end{pmatrix}_{n_i \times n_i}$$

$$= D_i \alpha + I_{n_i},$$

where

$$D_i = R_i(\boldsymbol{o}_1) - I_{n_i} = \begin{pmatrix} 0 & 1 & 1 & \cdots & 1 \\ 1 & 0 & 1 & \cdots & 1 \\ 1 & \vdots & \vdots & \ddots & 1 \\ 1 & 1 & \cdots & 1 & 0 \end{pmatrix}_{n_i \times n_i} = \frac{\partial R_i(\alpha)}{\partial \alpha} \tag{5.1}$$

and $\boldsymbol{o}_1 = 1$. (Recall that in the notation of the previous chapter, if the dimension of α is s_i, then $\boldsymbol{o}_j$ is defined to be a unit vector of length s_i with a one in the jth position. For the exchangeable structure, the dimension of α is one, so that $\boldsymbol{o}_1 = 1$.)

The exchangeable structure can also be expressed in the following simple form:

$$R_i(\alpha) = (1 - \alpha)I_{n_i} + \alpha J_{n_i}, \tag{5.2}$$

where $J_{n_i} = R_i(\boldsymbol{o}_1)$ is an $n_i \times n_i$ matrix of ones. In addition, J_{n_i} can be expressed as $\boldsymbol{e}_i \boldsymbol{e}_i'$, where $\boldsymbol{e}_i$ is an $n_i \times 1$ vector of ones.

Using Equation (5.2) it is straightforward to directly obtain the following expressions for the inverse and first derivative of the inverse of the exchangeable structure:

$$R_i^{-1}(\alpha) = \frac{1}{(1 - \alpha)}I_{n_i} - \frac{\alpha}{(1 - \alpha)\{1 + (n_i - 1)\alpha\}}J_{n_i} \tag{5.3}$$

and

$$\frac{\partial}{\partial \alpha}\left\{ R_i^{-1}(\alpha) \right\} = \frac{1}{(1 - \alpha)^2}\left\{ I_{n_i} - b_i(\alpha)J_{n_i} \right\}, \tag{5.4}$$

where

$$b_i(\alpha) = \frac{1 + (n_i - 1)\alpha^2}{\{1 + (n_i - 1)\alpha\}^2}. \tag{5.5}$$

For the exchangeable structure and for unbalanced data, Shults (1996) substituted Equation (5.4) into the stage one estimating equation (3.40) and simplified, to obtain the following estimating equation:

$$\sum_{i:n_i>1} Z_i(\widehat{\boldsymbol{\beta}})' Z_i(\widehat{\boldsymbol{\beta}}) - \sum_{i:n_i>1} \frac{1 + \alpha^2(n_i - 1)}{\{1 + \alpha(n_i - 1)\}^2}\{Z_i(\widehat{\boldsymbol{\beta}})' \boldsymbol{e}_i\}^2 = 0. \tag{5.6}$$

There will always exist a unique solution to Equation (5.6) for α in the feasible interval $(-1/(n_{max} - 1), 1)$, as established in a proof for the exchangeable structure in Shults (1996) and later in the more general Theorem 4.3 of Xie et al. (2010). Chaganty (1997) obtained an explicit solution to Equation (5.6) for the special case of balanced data. However, for unbalanced data there is no explicit solution, so that an iterative approach such as bisection must be used to obtain a solution to Equation (5.6) in the feasible interval for α.

In contrast to the stage one QLS estimating equation for the exchangeable structure, there is an explicit solution $\widehat{\alpha}_{QLS}$ to the stage two estimating equation (3.41). Direct substitution of Equation (5.4) into Equation (3.41) yields the following estimating equation:

$$\sum_{i=1}^{m} trace\left[\{I_{n_i} - b_i(\widehat{\alpha}_{QONE})J_{n_i}\}\{(1 - \alpha)I_{n_i} + \alpha J_{n_i}\}\right] = 0. \tag{5.7}$$

Simplification and some algebra can then be used to solve Equation (5.7) for α, in order to obtain the following stage two estimator $\widehat{\alpha}_{QLS}$ that was provided in (C.3) of Shults and Morrow (2002) (for $s_i = n_i$ and when (C.3) is calculated over all i, that is, when $g_i = 1$ for all i in the notation of Shults and Morrow (2002)):

$$\sum_{i:n_i>1} \frac{n_i(n_i-1)\,\widehat{\alpha}_{QONE}\,\{\widehat{\alpha}_{QONE}(n_i-2)+2\}}{\{1+\widehat{\alpha}_{QONE}(n_i-1)\}^2} \Big/ \sum_{i:n_i>1} \frac{n_i(n_i-1)\left\{1+\widehat{\alpha}_{QONE}^2(n_i-1)\right\}}{\{1+\widehat{\alpha}_{QONE}(n_i-1)\}^2}. \tag{5.8}$$

The estimator provided in (5.8) can also be obtained using Equation (4.18).

5.2.2 Demonstration of Implementation of the Exchangeable Structure for QLS

Next we demonstrate implementation of QLS by replicating the longitudinal study in obesity example provided in Section 2.4 of Chapter 2; as in that section, we do not display the entire iteration log and after the first example, we only display part of the output. We start by fitting in Stata the exchangeable structure with model-based covariance matrix.

```
. use random_small
. sort id month
. by id: gen lag = month - month[_n-1] if _n>1
. gen change = bmiz - basebmiz
. gen month2 = month^2
. keep if month<=36

. xtqls change month month2, i(id) t(month)   c(exc) vce(model) f(gau)

Iteration 2: tolerance = 1.780e-16

GEE population-averaged model          Number of obs      =    466
Group and time vars:  id __00000S      Number of groups   =    100
Link:     identity        Obs per  group:            min =      2
Family:    Gaussian                              avg =    4.7
Correlation:   fixed (specified)                 max =      6
```

```
Wald chi2(2)        =      15.47
Scale parameter:  .8112303              Prob > chi2     =    0.0004
```

change	Coef.	Std. Err.	z	P>|z|	[95% Conf. Interval]
month	.0168735	.0076202	2.21	0.027	.0019383	.0318088
month2	-.0006488	.0002136	-3.04	0.002	-.0010674	-.0002302
_cons	.8715136	.0876538	9.94	0.000	.6997154	1.043312

The results that we obtained above for QLS were almost identical to those obtained for GEE for an exchangeable working structure and model-based covariance matrix in Section 2.4 of Chapter 2. However, note that the output above indicates that the correlation is "fixed (specified)"; this is because **xtqls** calls up the **xtgee** procedure within each iteration and forces it to solve the GEE estimating equation evaluated at the current QLS estimate of α. Later in this chapter we will provide the algorithm that commands such as **xtqls** implement, in order to make use of existing software for GEE.

To present our results in a concise fashion, we next extract and display the estimate of α, rather than print the entire correlation structure.

```
. matrix working_structure = e(R)
. scalar alpha = working_structure[2,1]
. scalar list alpha
```

```
     alpha =    .7139654
```

The GEE estimate of α was ≈ 0.761, while the QLS estimate was $\hat{\alpha}_{QLS} \approx 0.714$.

Next, we fit the same model with the sandwich covariance matrix specified in the option vce(robust), to again obtain results that are very similar to those obtained earlier for GEE. (See Section 2.3.7 for a discussion of the sandwich covariance matrix.)

```
. xtqls change month month2, i(id)  t(month)  c(exc)  vce(robust)  f(gau)
```

(Std. Err. adjusted for clustering on id)

change	Coef.	Semirobust Std. Err.	z	P>|z|	[95% Conf. Interval]
month	.0168735	.0090067	1.87	0.061	-.0007793	.0345264
month2	-.0006488	.0002209	-2.94	0.003	-.0010818	-.0002158
_cons	.8715136	.073273	11.89	0.000	.7279012	1.015126

Implementation of the exchangeable structure was almost the same for QLS versus GEE, for this particular example. In practice, we have not identified any important differences between QLS and GEE for the exchangeable structure.

5.3   The Tri-Diagonal Correlation Structure

5.3.1   Solutions to the QLS Stage One and Stage Two Estimating Equations for α

The tri-diagonal structure defined in structure (2.13) is not widely used in analysis of medical studies; however, it is one of the basic structures considered in Liang and Zeger (1986) and it is available in the major software packages for GEE.

The $n_i \times n_i$ tri-diagonal structure (2.13) is a linear correlation structure because it can be expressed in the form (4.12) for a linear correlation structure:

$$
\begin{aligned}
R_i(\alpha) \;=\;& \{R_i(\boldsymbol{o}_1) - I_{n_i}\}\alpha + I_{n_i} \\[2mm]
=\;& \left\{
\begin{pmatrix}
1 & 1 & 0 & \cdots & 0 \\
1 & 1 & 1 & \cdots & 0 \\
0 & 1 & 1 & \cdots & \vdots \\
\vdots & \vdots & \vdots & \ddots & 1 \\
0 & 0 & \cdots & 1 & 1
\end{pmatrix}_{n_i \times n_i}
-
\begin{pmatrix}
1 & 0 & 0 & \cdots & 0 \\
0 & 1 & 0 & \cdots & 0 \\
0 & 0 & 1 & \cdots & 0 \\
\vdots & \vdots & \vdots & \ddots & \vdots \\
0 & 0 & 0 & \cdots & 1
\end{pmatrix}_{n_i \times n_i}
\right\}\alpha \\[2mm]
&+
\begin{pmatrix}
1 & 0 & 0 & \cdots & 0 \\
0 & 1 & 0 & \cdots & 0 \\
0 & 0 & 1 & \cdots & 0 \\
\vdots & \vdots & \vdots & \ddots & \vdots \\
0 & 0 & 0 & \cdots & 1
\end{pmatrix}_{n_i \times n_i} \\[2mm]
=\;& D_i\alpha + I_{n_i},
\end{aligned}
$$

where

$$
D_i = R_i(\boldsymbol{o}_1) - I_{n_i} =
\begin{pmatrix}
0 & 1 & 0 & \cdots & 0 \\
1 & 0 & 1 & \cdots & 0 \\
0 & 1 & 0 & \cdots & \vdots \\
\vdots & \vdots & \vdots & \ddots & 1 \\
0 & 0 & \cdots & 1 & 0
\end{pmatrix}_{n_i \times n_i}
= \frac{\partial R_i(\alpha)}{\partial \alpha}. \tag{5.9}
$$

For the tri-diagonal structure and unbalanced data, there will always be a feasible solution to the stage one estimating equation (3.40) for α, as established in a proof for the tri-diagonal structure by Shults (1996) and later in the more general Theorem 4.3 by Xie et al. (2010). However, simplifying and solving the stage one and two estimating equations (3.40) and (3.41) for the tri-diagonal structure are complicated by the fact that there is no simple expression for the inverse of a tri-diagonal structure; hence, it is not possible to obtain a simple expression for the *derivative* of the inverse that is involved in Equations (3.40) and (3.41). The various software implementations of QLS therefore make use of the following expression for the derivative of the inverse of a matrix:

$$
\frac{\partial R_i^{-1}(\alpha)}{\partial \alpha} = -R_i^{-1}(\alpha)\frac{\partial R_i(\alpha)}{\partial \alpha}R_i^{-1}(\alpha), \tag{5.10}
$$

where $\frac{\partial R_i(\alpha)}{\partial \alpha} = D_i$ in Equation (5.9) for the tri-diagonal structure.

Substitution of Equation (5.10) into Equation (3.40) then yields the following equivalent expression for the QLS stage one estimating equation for α:

$$\sum_{i=1}^{m} Z_i'(\widehat{\boldsymbol{\beta}})R_i^{-1}(\alpha)\frac{\partial R_i(\alpha)}{\partial \alpha}R_i^{-1}(\alpha)Z_i(\widehat{\boldsymbol{\beta}}) = 0. \tag{5.11}$$

Estimating equation (5.11) can be implemented when there is no simple expression for the inverse of a particular structure. To solve Equation (5.11), current programs for QLS (e.g. **xtqls** in Stata) use an iterative approach such as bisection that first constructs the structures $R_i(\alpha)$ at the current estimate of α and next obtains their inverses using a built-in command such as **syminv** in Stata, which obtains the inverse of a symmetric matrix.

Next, to obtain a solution to the *stage two estimating equation* (3.41) for the tri-diagonal structure, making use of Equations (5.9) and (5.10) we note that

$$\sum_{i=1}^{m} trace \left\{ \frac{\partial R_i^{-1}(\alpha)}{\partial \alpha}\bigg|_{\widehat{\alpha}_{QONE}} R_i(\alpha) \right\} = 0$$

$$\Longrightarrow \sum_{i=1}^{m} trace \left\{ R_i^{-1}(\widehat{\alpha}_{QONE})D_i R_i^{-1}(\widehat{\alpha}_{QONE})(D_i\alpha + I_{n_i}) \right\} = 0. \tag{5.12}$$

Estimating equation (5.12) has the following solution

$$\widehat{\alpha}_{QLS} = -\frac{\sum_{i=1}^{m} trace\,(H_i)}{\sum_{i=1}^{m} trace\,(H_i D_i)}, \tag{5.13}$$

where

$$H_i = R_i^{-1}(\widehat{\alpha}_{QONE})D_i R_i^{-1}(\widehat{\alpha}_{QONE}). \tag{5.14}$$

Because no closed-form expression is available for the inverse of the tri-diagonal structure, no further simplification is possible for Equation (5.13), as it was in Equation (5.8) for the exchangeable structure.

By Theorem 3.2 of Chaganty and Shults (1999) we know that $\widehat{\alpha}_{QLS}$ will be consistent for α, which means that asymptotically it will be feasible. However, as is true for GEE, there is no guarantee that feasible estimates will be obtained in practice.

5.3.2 Demonstration of Implementation of the Tri-Diagonal Structure for QLS

Next we demonstrate implementation of QLS by replicating the example provided in Section 2.4 of Chapter 2. We start by fitting the tri-diagonal structure with the sandwich covariance matrix. (*Capture* at the start of the command forces Stata to continue to run, even if convergence is not achieved.)

```
.capture xtqls2 change month month2, i(id) t(month)  c(sta 1) vce(robust) f(gau)

Iteration 1: tolerance = .0265037
Iteration 2: tolerance = 0
```

```
GEE population-averaged model              Number of obs   =  466
Group and time vars:     id __00001C       Number of groups=  100
Link:      identity      Obs per  group:     min =  2
Family:     Gaussian                         avg =  4.7
Correlation:           fixed (specified)     max =  6
                                   Wald chi2(2)  =  0.41
Scale parameter:      .8206121     Prob > chi2   = 0.8150
```

(Std. Err. adjusted for clustering on id)

```
-------------------------------------------------------------------------
              |              Semirobust
    change |    Coef.    Std. Err.    z    P>|z|    [95% Conf. Interval]
-------------+-----------------------------------------------------------
     month |   .0079766   .0169251   0.47   0.637   -.025196    .0411492
    month2 |  -.0002628   .0004687  -0.56   0.575   -.0011815   .0006559
     _cons |   .8168434   .0988331   8.26   0.000    .6231341  1.010553
-------------------------------------------------------------------------
```

convergence not achieved r(430);

It is interesting to note that when we implemented the tri-diagonal structure for GEE in Section 2.4, we also obtained a warning that "convergence was not achieved." We explored the reason for the failure and found that it was due to a final estimated correlation matrix that was not positive definite. We showed that the feasible interval for the tri-diagonal structure when $n_{max} = 6$, is approximately $(-0.554958, 0.554958)$. Next we obtain the QLS estimate of α.

```
. matrix working_structure = e(R)

. *Next, extract the estimate of alpha and display it:
. scalar alpha = working_structure[2,1]

. scalar list alpha
      alpha =   .56244927
```

Our QLS estimate of $\widehat{\alpha}_{QLS}$ is 0.5624, which is just barely outside the feasible interval for α. Therefore, like the GEE estimate (which was 0.8583), the QLS estimate is also infeasible for this structure. (In contrast, the QLS stage one estimate of α is guaranteed to be feasible. For this example, the stage one estimate was 0.4184.)

5.4   The AR(1) Structure for Analysis of (Planned) Equally Spaced Longitudinal Data

5.4.1   Solutions to the QLS Stage One and Stage Two Estimating Equations for α

The AR(1) structure was described and defined in Section 2.3.3 of Chapter 2. The correlations for this structure only depend on separation in order of measurement, so that, for example, the correlation between adjacent measurements on a subject would be assumed to be equal across all visits and subjects. For example, if $\alpha = 0.50$, then the assumed correlation between any measurements separated by one visit (e.g., (4th and 5th or 1st and 2nd) would be 0.50, while the correlation between any measurements separated by two visits (e.g., (4th and 6th or 1st and 3rd) would be

$0.50^2 = 0.25$. As a result, this structure is perhaps most appropriate for longitudinal data that are equally spaced (or approximately equally spaced) in time.

With respect to the patterns of timings described in Section 5.1, the AR(1) structure might be most plausible when the study design calls for equally spaced measurements and the data are complete or subjects do not return once they drop out of the study. However, the AR(1) structure can be applied in other situations, for example, in studies with planned measurement times that are unequal, but that are the same for all subjects. For example, in a study with measurement times at baseline and 3 and 12 months post baseline, an AR(1) structure would be plausible if we anticipate that the correlation between measurements only depends on the visit number, and not on the actual spacing between measurements. In other words, the AR(1) structure will be appropriate for this example if the correlation between baseline and 3 months (visits 1 and 2, with a time-gap of 3 months) is expected to be the same as the correlation between 3 and 12 months (visits 2 and 3, with a time gap of 9 months).

In order to streamline our discussion of the development of QLS in Chapter 3, we presented our results for the AR(1) structure; however, we could just as easily have featured another structure. One advantage of the AR(1) structure is that it has simple explicit solutions to the stage one and two QLS estimating equations, that were provided in Equations (3.31) and (3.37), respectively.

5.4.2 *Demonstration of Implementation of the AR(1) Structure for QLS*

Data analysis examples for QLS and the AR(1) structure were provided in Section 3.6 in Chapter 3. The results for QLS were almost identical to those that were obtained for GEE in Chapter 2. Neither approach identified a significant change in BMI z-score over time, which was *not consistent* with our prior knowledge that children tend to initially gain weight following renal transplant.

Here we provide a caution to the reader regarding datasets that contain a variable number of measurements per subject (or cluster). There is one potential major difference between the implementations of QLS that are currently available in Stata, SAS, R, and MATLAB, versus some software programs that are currently available for GEE. The software for QLS has been programmed so that subjects with only one measurement *will not* be dropped from the analysis; rather, subjects with at least two measurements will provide information with respect to estimation of α, while all subjects will provide information with respect to β. We demonstrate the difference here. Continuing with the previous example for the tri-diagonal structure that was presented in Section 5.6.3, we start by dropping all but the first measurement on subjects with identification number < 50.

```
. sort id month
. by id: drop if id < 50 & _n>1
```

```
(163 observations deleted)
```

Next, we use the command **xtgee** in Stata to fit our prior model with GEE, but for the reduced dataset.

```
. xtgee change month month2, i(id) t(month) force c(AR 1)

note: some groups have fewer than 2 observations
        not possible to estimate correlations for those groups
        41 groups omitted from estimation

Iteration 4: tolerance = 2.527e-07

GEE population-averaged model          Number of obs      = 262
Group and time vars:   id month        Number of groups   = 59
Link:      identity       Obs per group:        min = 2
Family:      Gaussian                            avg = 4.4
Correlation: AR(1)                               max = 6
                                     Wald chi2(2)   =    7.22
Scale parameter:   .7648645            Prob > chi2  = 0.0270
```

change	Coef.	Std. Err.	z	P>|z|	[95% Conf. Interval]	
month	.0199844	.0095349	2.10	0.036	.0012964	.0386725
month2	-.0006438	.0002495	-2.58	0.010	-.0011327	-.0001549
_cons	.8442925	.1151514	7.33	0.000	.6185999	1.069985

The note that "41 groups omitted from estimation" indicates that 41 subjects were dropped from the analysis because they had only one measurement. Therefore, only 59 subjects remain in the analysis. Next, we fit QLS for the same dataset.

```
. xtqls change month month2, i(id) t(month)  c(AR 1) vce(robust) f(gau)

Iteration 1: tolerance = .02663888
Iteration 2: tolerance = 1.277e-16

GEE population-averaged model          Number of obs      = 303
Group and time vars: id __00000S       Number of groups   = 100
Link:      identity       Obs per group:        min = 1
Family:      Gaussian
                                                avg = 3.0
Correlation:   fixed (specified)                max = 6
                                     Wald chi2(2)   =    8.22
Scale parameter:   .7615011            Prob > chi2  = 0.0164

                                 (Std. Err. adjusted for clustering on id)
```

change	Coef.	Semirobust Std. Err.	z	P>|z|	[95% Conf. Interval]	
month	.0239364	.011937	2.01	0.045	.0005403	.0473326
month2	-.0007051	.000283	-2.49	0.013	-.0012598	-.0001505
_cons	.7394216	.0779732	9.48	0.000	.586597	.8922463

We now see that the number of subjects is 100, with a number of measurements per subject that ranged from 1 to 6, so that no subjects were dropped from the analysis. When we next display the estimated correlation matrix we see that the largest working structure, for $n_{max} = 6$, is displayed.

```
. xtcorr
```

Estimated within-id correlation matrix R:

```
        c1       c2       c3       c4       c5       c6
r1   1.0000
r2   0.8559   1.0000
r3   0.7325   0.8559   1.0000
r4   0.6270   0.7325   0.8559   1.0000
r5   0.5366   0.6270   0.7325   0.8559   1.0000
r6   0.4593   0.5366   0.6270   0.7325   0.8559   1.0000
```

The results did not differ greatly between the larger and smaller datasets, but the QLS results are preferable because they are based on a larger dataset. We therefore caution the reader to check whether subjects with one measurement are being excluded from the analysis, for the particular software package you are using for GEE.

5.5   The Markov Structure for Analysis of Unequally Spaced Longitudinal Data

5.5.1   Solutions to the QLS Stage One and Stage Two Estimating Equations for α

The Markov correlation structure was not presented in Chapter 2, because it is not currently available in the major software packages that implement GEE. We therefore present it here.

For the Markov structure, the correlation between two measurements collected on subject i at times t_{ij} and t_{ik} is given by $Corr(y_{ij}, y_{ik}) = \alpha^{|t_{ij} - t_{ik}|}$. The Markov structure therefore generalizes the AR(1) structure so that the correlations no longer depend only on differences in visit numbers, but on their separation in time. If the planned timings are equally spaced, then the estimated correlation matrices will be identical for the Markov and AR(1) structures, so that the Markov structure includes the AR(1) structure as a special case.

The Markov structure can be applied for any of the patterns of timings of measurements that were described in Section 5.1. It might also be applied in analysis of data with equal spacing of consecutive measurements on subjects, if the actual dates of measurements are taken into account. For example, the AR(1) structure might be implemented in the first analysis of complete data from a study with measurements collected at planned measurement times of 0, 6, and 12 months. The Markov structure might then be implemented if we base our analysis on the actual dates of measurements, that yield timings that are approximately equal (but not identical) to 6 and 12 months post baseline. In general, the Markov structure might be uniquely appropriate for studies with unequal and irregular spacing of measurements, as in Preisser et al. (2003a).

An $n_i \times n_i$ Markov structure has the following form:

$$
\begin{pmatrix}
1 & \alpha^{t_{i2}-t_{i1}} & \alpha^{t_{i3}-t_{i1}} & \cdots & \alpha^{t_{in_i}-t_{i1}} \\
\alpha^{t_{i2}-t_{i1}} & 1 & \alpha^{t_{i3}-t_{i2}} & \cdots & \alpha^{t_{in_i}-t_{i2}} \\
\alpha^{t_{i3}-t_{i1}} & \alpha^{t_{i3}-t_{i2}} & 1 & \cdots & \alpha^{t_{in_i}-t_{i3}} \\
t_{i4}-t_{i1} & & & & \\
\vdots & \vdots & \vdots & \ddots & \vdots \\
\alpha^{t_{in_i}-t_{i1}} & \alpha^{t_{in_i}-t_{i2}} & \alpha^{t_{in_i}-t_{i3}} & \cdots & 1
\end{pmatrix}_{n_i \times n_i} .
\tag{5.15}
$$

The feasible interval for this structure is $\alpha \in (0,1)$. An additional requirement is that $t_{ij} - t_{ij-1} \geq 1$ for all i and $j = 2, \ldots, n_i$. This means that the time scale must be such that the time between any two consecutive measurements on a subject is at least one. If necessary, the data can be easily re-scaled to satisfy this condition. For example, time in months could be re-scaled as time in days in order to ensure that all time gaps are at least one.

The AR(1) structure may be appropriate for unbalanced data from a study with a set of *planned* timings that are equally spaced. The Markov structure is a generalization of the AR(1) structure that might be plausible in a longitudinal study when the timings between measurements are not the same for all subjects; in this situation, an AR(1) structure might not be appropriate. For example, if the separation in time between the first and second measurements is 1 week for some subjects versus several months for others, then it might be unreasonable to assume that the correlation between the first and second measurements is the same for all subjects. The Markov correlation structure is available in the MATLAB, SAS, Stata, and R software (Ratcliffe and Shults, 2008; Kim and Shults, 2010; Shults et al., 2007; Xie and Shults, 2009) for QLS.

For the Markov structure and unbalanced data, Shults (1996) obtained the QLS stage one estimating equation for α and proved that it will always contain a unique solution for $\alpha \in (0,1)$:

$$
\sum_{i=1}^{m} \sum_{j=2}^{n_i} \frac{\delta_{ij} \alpha^{\delta_{ij}} \left[\alpha^{2\delta_{ij}} z_{ij} z_{i,j-1} - \alpha^{\delta_{ij}} \left(z_{ij}^2 + z_{i,j-1}^2 \right) + z_{ij} z_{i,j-1} \right]}{(1 - \alpha^{2\delta_{ij}})^2} = 0,
\tag{5.16}
$$

where $\delta_{ij} = |t_{ij} - t_{i,j-1}|$.

The stage two estimating equation for the Markov structure is given by

$$
\sum_{i=1}^{m} \sum_{j=2}^{n_i} \frac{2\delta_{ij} \widehat{\alpha}_{QLSONE}^{2\delta_{ij}-1} - \alpha^{\delta_{ij}} \delta_{ij} \left[\widehat{\alpha}_{QLSONE}^{\delta_{ij}-1} + \widehat{\alpha}_{QLSONE}^{3\delta_{ij}-1} \right]}{(1 - \widehat{\alpha}_{QLSONE}^{2\delta_{ij}})^2} = 0.
\tag{5.17}
$$

Chaganty and Shults (1999) proved that there will always be a unique solution to Equation (5.17) for $\alpha \in (0,1)$. It is straightforward to show that if the $\delta_{ij} = 1$ for all i and j, in which case the Markov structure reduces to AR(1), there are explicit solutions to Equations (5.16) and (5.17) that are identical to those provided for the AR(1) structure in (3.31) and (3.37), respectively.

5.5.2 *Demonstration of Implementation of the Markov Structure for QLS*

Next, we will recreate the full dataset from the longitudinal study of obesity that we have been considering in this chapter, and will fit the Markov correlation structure with a model based covariance matrix:

```
. use random_small, clear
. sort id month
. by id: gen lag = month - month[_n-1] if _n>1
. gen change = bmiz - basebmiz
. gen month2 = month^2
. keep if month<=36

. xtqls change month month2, i(id) t(month) vce(model) f(gau) c(Markov)

Iteration 1: tolerance = .11779999
Iteration 2: tolerance = 6.743e-17

GEE population-averaged model            Number of obs   = 466
Group and time vars: id month      Number of groups = 100
Link:        identity        Obs per group:        min = 2
Family:      Gaussian
                                                   avg = 4.7
Correlation:        fixed (specified)              max = 6
Wald chi2(2)      =      22.14    Scale parameter:    .8260266
Prob > chi2       =      0.0000
```

change	Coef.	Std. Err.	z	P>|z|	[95% Conf. Interval]
month	.042706	.0092227	4.63	0.000	.0246298	.0607822
month2	-.0011522	.0002524	-4.56	0.000	-.001647	-.0006574
_cons	.6464564	.0953885	6.78	0.000	.4594984	.8334145

The estimates of the regression coefficients differ between the AR(1) and Markov correlation structures, for example the coefficient for *month* is ≈ 0.043 for the Markov structure, versus ≈ 0.024 for the AR(1) structure. In addition, the time effects are only significant for the Markov structure. We also obtain significant time effects when the sandwich covariance matrix is specified:

```
. xtqls change month month2, i(id) t(month) vce(robust) f(gau) c(Markov)

                                (Std. Err. adjusted for clustering on id)
```

		Semirobust				
change	Coef.	Std. Err.	z	P>|z|	[95% Conf. Interval]
month	.042706	.0089966	4.75	0.000	.0250729	.0603391
month2	-.0011522	.0002231	-5.16	0.000	-.0015895	-.0007149
_cons	.6464564	.0719742	8.98	0.000	.5053896	.7875232

Next we display the estimated Markov correlation structure:

```
. xtcorr
```

```
Estimated within-id correlation matrix R:
```

```
         c1      c2      c3      c4      c5      c6
r1   1.0000
r2   0.9272  1.0000
r3   0.8277  0.8928  1.0000
r4   0.6597  0.7116  0.7970  1.0000
r5   0.4191  0.4520  0.5063  0.6352  1.0000
r6   0.2662  0.2871  0.3216  0.4035  0.6352  1.0000 .
```

In contrast to the AR(1) structure, the off-diagonal elements are not equal for the Markov structure. This is because the correlations between consecutive measurements are displayed on the off-diagonal, and the gap times between the timings of consecutive planned measurements are not identical. Note also that in this study, subjects had between two and six measurements, and all subjects had timings that were a subset of a common set of measurement times $\{1, 3, 6, 12, 24, 36\}$. If a subject only had measurements taken at times $\{1, 3, 36\}$, that is, at the 1st, 2nd, and 6th measurement occasions, then her working correlation structure would be obtained by selecting the 1st, 2nd, and 6th rows and columns of the above estimated correlation matrix, and could be represented as follows:

```
         c1      c2      c6
r1   1.0000
r2   0.9272  1.0000
r6   0.2662  0.2871  1.0000
```

That the time effects differ markedly for the Markov structure both with respect to the estimated values and their significance, for example, the time effects were significant only for the Markov structure, suggests that we will need to carefully compare the fit of competing models for this analysis. We will return to this example again in Exercise 8.6.

5.5.3   Generalized Markov Structure

The generalized Markov structure further generalizes the Markov structure in (5.15), so that element (j,k) $(j < k)$ of the structure is $\alpha^{(t_k^{\gamma} - t_j^{\gamma})/\gamma}$ $(\gamma > 0)$. As described in Shults and Chaganty (1998), if $\gamma \approx 0$, the correlation between two measurements on a subject that are separated by w time units will be greater if the measurements were collected later in the study, rather than earlier. For example, if $\gamma = 0.1$ and $\alpha = 0.60$, the correlation between adjacent measurements separated by 1 week will be $0.60^{(2^{0.10} - 1^{0.10})/0.10} = 0.69$, if they were measured at weeks 1 and 2, but will be $0.60^{(22^{0.10} - 21^{0.10})/0.10} = 0.97$, if they were measured at weeks 21 and 22. This structure is therefore plausible when we anticipate that the correlation between two measurements that are separated by k time-units will be greater later in the study, rather than earlier. The parameter γ can also be viewed as a dampening parameter, that dampens the decay in correlation with increasing separation in time, relative to the Markov structure. For example, if $\alpha = 0.50$ and $(t_1, t_2) = (1, 9)$, $\mathrm{corr}(Y_{i1}, Y_{i2}) =$

$0.50^{(9^1 - 1^1)/1} = 0.002$ for the Markov structure, but will be $0.50^{(9^{0.1} - 1^{0.1})/0.1} = 0.166$ for the generalized Markov structure (with $\gamma = 0.10$).

Due to some problems with estimating the parameter γ, this structure is not currently available in the general software packages that implement QLS. It was implemented using a grid search in Shults and Chaganty (1998).

5.6   The Unstructured Matrix for Analysis of Balanced Data

The unstructured matrix was defined in (2.14) and demonstrated for GEE in Section 2.4.

With respect to the characteristics described in Section 5.1, the unstructured matrix might be most applicable for studies with a common set of measurement times, whether they are equally or unequally spaced in time. For example, consider a study that planned for measurements to be collected at 0, 6, 12, and 18 months post baseline (so that the study design called for equally spaced measurements at fixed time points). If the second visit was missing on all subjects, then the resultant measurements would be available at 0, 12, and 18 months post baseline. The unstructured matrix would be applicable for these unequally spaced measurements. In addition, *when GEE is implemented in the analysis*, the unstructured matrix can be applied when the number of measurements varies between subjects. For example, in the previous example if some subjects also dropped out of the study but did not return after drop out, then the unstructured matrix might be appropriate. It is also possible to implement GEE for an unstructured matrix when subjects miss visits prior to their final visit, as discussed in Section 2.3.5. However, as noted in Section 2.3.5, we do suspect that greater imbalance in the data can lead to greater instability in the GEE estimation procedure.

Unlike GEE, QLS requires an equal number of measurements per subject for implementation of the unstructured matrix. The key question with respect to whether the unstructured matrix is plausible for a QLS analysis is whether it is reasonable to assume that the correlation between any two particular measurement occasions is the same for all subjects. For example, suppose that the first, second, and third measurements were collected at baseline, 6, and 12 months on some subjects, and at baseline, 7, and 13 months on others. The unstructured matrix will be appropriate if it is reasonable to assume that the correlation between the first and second measurements on a subject is the same for all subjects, that is, is the same for measurements taken at (baseline, 6 months) versus (baseline, 7 months). In addition, is it reasonable to assume that the correlation between the second and third measurements, or first and third, is the same for all subjects, even though the temporal spacing between these measurements is not exactly the same for all subjects? If so, then we can create a timing variable that represents the order of measurement (and therefore takes value 1, 2, 3) and use this as the timing variable in our QLS analysis.

In Section 4.7 we directly implemented QLS for a 3×3 unstructured matrix by obtaining explicit solutions to the stage one and two estimating equations for that particular structure. However, in general, implementation of a completely unstructured matrix for QLS is difficult because solving the estimating equations for the correlation parameters typically involves searching for solutions within the feasible region that yields positive definite correlation matrices. There is no simple form for either the inverse or the feasible region, for an $n \times n$ unstructured matrix. However, it is possible to implement an unstructured matrix for QLS, for data that are balanced, so that the timing of measurements and the number of measurements is the same for all subjects.

5.6.1 Obtaining a Solution to the Stage One Estimating Equation for the Unstructured Matrix

The $n \times n$ unstructured matrix (2.14) is a linear correlation structure because it can be expressed in the form (4.12) for a linear correlation structure:

$$
R(\boldsymbol{\alpha}) = \sum_{j=1}^{n(n-1)/2} (R(\boldsymbol{o}_j) - I_n) \, \boldsymbol{\alpha}[j] + I_n
$$

$$
= \sum_{j=1}^{n(n-1)/2} \boldsymbol{D}_j \boldsymbol{\alpha}[j] + I_{n_i},
$$

where $\boldsymbol{D}_j = \frac{\partial R(\boldsymbol{\alpha})}{\partial \boldsymbol{\alpha}[j]}$ and $\boldsymbol{\alpha} = (\alpha_{12}, .., \alpha_{1n}, \alpha_{23}, .., \alpha_{2n}, .., \alpha_{n-1n})'$.

For the unstructured matrix, Theorem 4.3 establishes that there will always be a feasible (positive definite) solution to the stage one estimating equation (3.40) for $\boldsymbol{\alpha}$. However, obtaining a solution to the stage one estimating equation is extremely difficult for the unstructured matrix, for several reasons. There is no simple expression for the inverse of the unstructured matrix. In addition, the number of correlation parameters involved in this structure increases as the cluster size increases. In general, there is also no simple expression for the region on which the correlation parameters yield a positive definite correlation matrix. The various software implementations of QLS have not included the unstructured matrix due to the complexity of its implementation for QLS. However, the Stata command **xtqls2** (Shults) does allow for implementation of the unstructured matrix for balanced data, using the approach described here.

Using results by Whittle (1958) and Olkin and Pratt (1958), Chaganty (1997) suggested an approach to solving the stage one estimating equation (3.40) for an $n \times n$ unstructured matrix. (See Example 4.4 in Chaganty (1997).) First, recall that element j, k of the GEE moment estimator for the unstructured matrix was provided in Equation (2.20). This estimator can also be expressed in the following matrix form:

$$
\widehat{R}_{GEE} = \frac{1}{(m-p)\widehat{\phi}_{GEE}} \left(\widehat{\boldsymbol{Z}} - diag(\widehat{\boldsymbol{Z}}) \right) + I_n,
$$

where

$$
\boldsymbol{Z} = \sum_{i=1}^{m} Z_i(\boldsymbol{\beta}) Z_i'(\boldsymbol{\beta}) \tag{5.18}
$$

and $Z_i(\boldsymbol{\beta}) = A_i^{-1/2}(\boldsymbol{\beta}) \, (\boldsymbol{Y}_i - \boldsymbol{\mu}_i)$.

As we shall see, the QLS estimator of the unstructured matrix also depends on $\mathbf{Z}$, which will be positive definite if $n_i = n \; \forall \; i$ and $m \geq n$, so that the number of measurements per subject is the same for all subjects, and the number of subjects is at least as great as the number of measurements per subject. Next, recall that the solution $\widehat{\boldsymbol{\alpha}}_{QLSONE}$ to the stage one estimating equation for $\boldsymbol{\alpha}$ minimizes the generalized error sum $Q^*(\alpha, \beta)$ that was defined in Equation (3.13). We can also write

$$Q^*(\alpha, \beta) \;=\; \sum_{i=1}^{m} Z_i'(\boldsymbol{\beta}) R^{-1}(\boldsymbol{\alpha}) Z_i(\boldsymbol{\beta}) \tag{5.19}$$

$$=\; trace(\mathbf{Z} R^{-1}(\boldsymbol{\alpha})) \tag{5.20}$$

where $R_i(\boldsymbol{\alpha}) = R(\boldsymbol{\alpha})$ for all i when the data are balanced. In other words, $trace(\mathbf{Z} R^{-1}(\widehat{\boldsymbol{\alpha}}_{QLSONE}))$ is the smallest possible value of $trace(\mathbf{Z} R^{-1}(\boldsymbol{\alpha}))$, if a suitable correlation matrix $R(\widehat{\boldsymbol{\alpha}}_{QLSONE})$ exists. That a suitable matrix will indeed always exist was proven by Whittle (1958, p. 234, Lemma 3) and Olkin and Pratt (1958).

Results provided in Olkin and Pratt (1958) establish that $R(\widehat{\boldsymbol{\alpha}}_{QLSONE})$ can be expressed as

$$R(\widehat{\boldsymbol{\alpha}}_{QLSONE}) = \Delta^{-1/2} \left(\Delta^{1/2} \mathbf{Z} \Delta^{1/2} \right)^{1/2} \Delta^{-1/2}, \tag{5.21}$$

where Δ is a diagonal matrix with positive elements that satisfies the fixed point equation

$$\Delta = diag \left(\Delta^{1/2} \mathbf{Z} \Delta^{1/2} \right)^{1/2}. \tag{5.22}$$

The matrix $R(\widehat{\boldsymbol{\alpha}}_{QLSONE})$ provided in Equation (5.21) is the solution to the equation

$$\mathbf{Z} = R(\widehat{\boldsymbol{\alpha}}_{QLSONE}) \Delta R(\widehat{\boldsymbol{\alpha}}_{QLSONE}). \tag{5.23}$$

To find $R(\widehat{\boldsymbol{\alpha}}_{QLSONE})$ for a particular value of $\mathbf{Z}$, we can proceed as follows. First we obtain Δ that satisfies Equation (5.22) by taking the following steps that were suggested in Chaganty (1997).

1. Choose starting values for the diagonal elements of Δ_1, for example, $\Delta_1 = diag(1, 1, \ldots, 1)$.

2. Next, for $k \geq 2$, define $\Delta_k = diag \left(\Delta_{k-1}^{1/2} \mathbf{Z} \Delta_{k-1}^{1/2} \right)^{1/2}$ and compare Δ_k with Δ_{k-1}. Stop the process and use Δ_k as the estimate of Δ when $\Delta_k \approx \Delta_{k-1}$, for example, when the absolute value of $e'(\Delta_k - \Delta_{k-1}) e$ is smaller than some prespecified tolerance value, where e is an $n \times 1$ vector of ones.

Chaganty (1997, p. 47) notes that "The proof that this fixed point iteration scheme converges to the unique solution of" Eq. (4.9) and related results will appear elsewhere." (Equation. (4.9) in Chaganty (1997, p. 47) is the same as Equation (5.22) above.) Therefore, there is no guarantee that the above algorithm will yield the desired solution. However, the algorithm does seem to work well in the examples that we have considered.

When there are only two measurements per subject, that is, when $n = 2$, the AR(1), tri-diagonal, exchangeable, and unstructured matrices are identical, so that the stage one (and two) QLS estimators should be equal for all structures in this special situation; in the exercises for this chapter, we directly solve Equation (5.23) and show that this is the case. In addition, in the appendix to this chapter, we consider the same example that we considered in Section 4.5 and demonstrate that this algorithm yields the same estimate that we obtained via direct solution of the stage one estimating equations for a 3×3 unstructured matrix.

5.6.2 Obtaining a Solution to the Stage Two Estimating Equation for the Unstructured Matrix

The unstructured matrix is a linear correlation structure. We could therefore obtain the expression for the stage two estimator for linear correlation structures that was provided in Equation (4.18). However, we could also use the following expression that was provided by Chaganty and Shults (1999, Section 3.1):

$$\widehat{R}_{QLS} = \widehat{R}_{QLSONE} \, diag(\hat{v}) \, \widehat{R}_{QLSONE}, \qquad (5.24)$$

where $\widehat{R}_{QLSONE}$ is the stage one estimate of the unstructured matrix; $\hat{v} = \widehat{R}_{QLSONE} \circ \widehat{R}_{QLSONE}^{-1} \mathbf{e}$; $A \circ B$ is the Hadamard product obtained by element-wise multiplication of A and B; and $\mathbf{e}$ is an $n \times 1$ vector of ones. It can be shown (see Exercise 5.7) that the estimators provided in Equations (4.18) and (5.24) are identical.

Theorem 4.4 of Xie et al. (2010) establishes that the stage two estimator (5.24) is consistent; however, there is no guarantee that it will be positive definite for finite samples. Chaganty and Shults (1999) provided the following estimator that could be used if the stage two estimator (5.24) is not positive definite:

$$\widehat{R}_{QLS} = diag(\widehat{\mathbf{Z}})^{-1/2} \widehat{\mathbf{Z}} diag(\widehat{\mathbf{Z}})^{-1/2}. \qquad (5.25)$$

The matrix (5.25) is easily proven to be positive definite, so that the final stage two estimator will be positive definite.

Because the procedure described here guarantees a positive-definite matrix, some might claim that QLS is preferable to GEE (because GEE does not offer this guarantee). However, the estimator (5.25) could also be implemented within the GEE estimation procedure, should the estimated correlation matrix for GEE fail to be positive definite. In other words, Xie et al. (2010) showed that the stage two QLS estimator for linear structures can be simply expressed as Equation (4.18), which according to Theorem 4.4 is guaranteed to be consistent but not necessarily positive definite. If the stage two estimator for the unstructured matrix (which can be expressed as Equation (4.18) or equivalently as Equation (5.24)) is not positive-definite, then we can implement the positive definite estimator (5.25). However, Equation (5.25) is not the stage two estimator that was obtained using the QLS procedure; that is, it is not necessarily a solution to the stage two estimating equation (3.41) for $\boldsymbol{\alpha}$; its implementation is therefore an ad-hoc correction that could also be applied for GEE.

In fact, Equation (5.25) was provided in provided in (7) of Park (1993), who obtained it as the maximum-likelihood estimator of the correlation matrix for multivariate Gaussian data when the standard deviations (scalar parameters in the context of GEE) are not forced to be equal at all measurement occasions. That Equation (5.25) allows for unequal standard deviations (in contrast to the estimator (2.20) that is usually implemented for GEE) can be seen by examining the form of the two estimators: Recall that estimator (2.20) has the following expression for element j,k of the estimated correlation matrix:

$$\widehat{R}_{GEE}[j,k] = \frac{\sum_{i=1}^{m}\widehat{z}_{ij}\widehat{z}_{ik}}{(m-p)\widehat{\phi}_{GEE}}. \tag{5.26}$$

This estimator is a function of the estimator $\widehat{\phi}_{GEE}$ of the scalar parameter ϕ, which is assumed to be constant at all measurement occasions. In contrast, element j,k of Equation (5.25) can be expressed as follows:

$$\widehat{R}_{GEE-Park}[j,k] = \frac{\sum_{i=1}^{m}\widehat{z}_{ij}\widehat{z}_{ik}}{m\widehat{\phi}_{j}\widehat{\phi}_{k}}, \tag{5.27}$$

where $\widehat{\phi}_{j}$ is the estimator of the scalar parameter ϕ_{j} at measurement occasion j. The estimator $\widehat{\phi}_{j}$ is the average of the estimated Pearson residuals at measurement occasion j:

$$\widehat{\phi}_{j} = \frac{\sum_{i=1}^{m}\widehat{z}_{ij}^{2}}{m}. \tag{5.28}$$

Therefore, while Equation (2.20) involves an estimator of the scalar parameter that is assumed to be constant over all measurement occasions (and that is obtained as the average of all estimated Pearson residuals), Equation (5.25) involves estimators of the scalar parameters that involve distinct estimators of the scalar parameter at measurement occasions j and k.

In Section 5.6.3 we provide an example for which GEE with an unstructured correlation matrix fails to converge for GEE. However, when the moment estimator (2.20) that is implemented by **xtgee** for the unstructured matrix is replaced with Equation (5.25), we demonstrate that GEE does converge. GEE will always converge for the estimator (5.25) because it is guaranteed to yield a positive-definite estimated working correlation structure *for complete data*. The estimator (5.27) could also be applied for unbalanced data; however, it is only guaranteed to be positive definite for complete data.

5.6.3   Demonstration of Implementation of the Unstructured Matrix for QLS

Next, we will continue with the example from Section 5.5.2. The algorithm for implementation of QLS for an unstructured matrix requires an equal number of measurements per subject. For demonstration, we retain only those subjects who had complete data up until their sixth measurement occasion.

```
.   use random_small, clear
.   sort id month
.   gen change = bmiz - basebmiz
.   gen month2 = month^2
```

```
. by id: keep if _N==6 & month[_N]==48
(483 observations deleted)
```

Next, we implement QLS with a sandwich covariance matrix using the user-written Stata command **xtqls2** (Shults).

```
xtqls2 change month month2, i(id) t(month) vce(robust) f(gau) c(uns)
We first checked that number of measurements per subject is
constant for the unstructured correlation structure.
```

```
The following matrix is Delta in (4.7) of Chaganty and Shults (1999).
Check that the diagonal elements are positive.
```

```
symmetric delta[6,6]
           c1          c2          c3          c4          c5          c6
r1   1.5342115
r2           0   1.7930011
r3           0           0   5.5426329
r4           0           0           0   .91255211
r5           0           0           0           0   2.2768766
r6           0           0           0           0           0   .67424495
```

```
The following matrices are Z and R Delta R, the left-hand side
and right-hand side of (4.7) of Chaganty and Shults.
Check that they are equal.
```

```
symmetric Z[6,6]
           r1          r2          r3          r4          r5          r6
r1   4.7338957
r2   4.4093486   5.6103945
r3    5.597415   6.5123493   8.0875643
r4   4.1208191   3.4896721   4.8087976   4.0838938
r5   4.6618328   4.2959748   5.8593305   4.5959756   5.5897906
r6   3.7336591   3.1645655   4.4191245   3.8272654   4.5799254   3.8834933
```

```
symmetric Z2[6,6]
           r1          r2          r3          r4          r5          r6
c1   4.7338957
c2   4.4093486   5.6103945
c3    5.597415   6.5123493   8.0875643
c4   4.1208191   3.4896721   4.8087976   4.0838938
c5   4.6618328   4.2959748   5.8593305   4.5959756   5.5897906
c6   3.7336591   3.1645655   4.4191245   3.8272654   4.5799254   3.8834933
```

```
The following are the RHS and LHS, respectively, of (4.9) in Chaganty & Shults
(1999). The following matrices must agree.
```

```
symmetric digammahalf[6,6]
           c1          c2          c3          c4          c5          c6
r1   1.5342115
r2           0   1.7930011
r3           0           0   5.5426329
r4           0           0           0   .91255211
r5           0           0           0           0   2.2768766
r6           0           0           0           0           0   .67424495
```

```
symmetric delta[6,6]
          c1          c2          c3          c4          c5          c6
r1  1.5342115
r2          0   1.7930011
r3          0           0   5.5426329
r4          0           0           0   .91255211
r5          0           0           0           0   2.2768766
r6          0           0           0           0           0   .67424495
```

Updated matrix in stage two:

```
symmetric __000002[6,6]
          r1          r2          r3          r4          r5          r6
c1          1
c2  .76755247           1
c3  .89110456   .94421858           1
c4  .90192423   .55530018   .7680212           1
c5  .88247915   .63434534   .82601389   .95257024           1
c6  .81965887   .52861928   .73010305   .93914562   .98195147           1
```

```
Iteration 1: tolerance = .19228394
Iteration 2: tolerance = 7.060e-17
```

```
GEE population-averaged model          Number of obs      =       48
Group and time vars:      id __00001C   Number of groups   =        8
Link:                        identity   Obs per group: min =        6
Family:                      Gaussian                 avg =      6.0
Correlation:        fixed (specified)                 max =        6
                                        Wald chi2(2)       =     2.73
Scale parameter:              .7153115   Prob > chi2       =   0.2558
```

(Std. Err. adjusted for clustering on id)

```
-----------------------------------------------------------------------
             |               Semirobust
    change   |     Coef.    Std. Err.      z    P>|z|   [95% Conf. Interval]
-------------+---------------------------------------------------------
     month   |   .0094224    .0057265    1.65   0.100   -.0018014   .0206462
    month2   |  -.0001285    .0000861   -1.49   0.136   -.0002973   .0000403
     _cons   |    .572525    .1400207    4.09   0.000    .2980896   .8469605
-----------------------------------------------------------------------
```

The **xtqls2** command first checked that the number of measurements was the same for all subjects. The command also displayed the estimated correlation matrix within each iteration of stage one. To save space, only the matrix for the final iteration (twenty) within stage one is shown above. Then, output was provided so that the user can check that Δ does have positive diagonal elements and satisfies Equation (5.22). In addition, output was provided so that the analyst can check that Δ and $R(\hat{\alpha}_{QLSONE})$ do indeed satisfy Equation (5.23). Then the updated stage two matrix was displayed, along with the output for the regression model. The final estimate of the correlation matrix suggests that there is a high degree of correlation among the repeated measurements on each subject. Furthermore, the pattern in the correlations does not seem to be compatible with either a Markov or AR(1) correlation structure.

If we fit the same model with GEE, we obtain a warning that the model failed to converge.

```
. xtgee change month month2, i(id) t(month) robust f(gau) c(uns)

Iteration 1: tolerance = .08688681
Iteration 2: tolerance = .07227717
Iteration 3: tolerance = .00469037
Iteration 4: tolerance = .0002487
Iteration 5: tolerance = 6.271e-06
Iteration 6: tolerance = 4.499e-08
```

GEE population-averaged model			Number of obs	=	48
Group and time vars:		id month	Number of groups	=	8
Link:		identity	Obs per group: min =		6
Family:		Gaussian	avg =		6.0
Correlation:		unstructured	max =		6
			Wald chi2(2)	=	3.79
Scale parameter:		.6853455	Prob > chi2	=	0.1500

(Std. Err. adjusted for clustering on id)

change	Coef.	Semirobust Std. Err.	z	P>|z|	[95% Conf. Interval]
month	.0154656	.0147738	1.05	0.295	-.0134905	.0444216
month2	-.0002995	.0002406	-1.24	0.213	-.0007712	.0001722
_cons	.6573898	.2799834	2.35	0.019	.1086324	1.206147

```
convergence not achieved
r(430);
```

This analysis demonstrates that QLS may converge for the unstructured correlation structure, while GEE fails to converge for the moment estimator of the unstructured matrix that is implemented in **xtgee** in Stata 13.0. However, there could be examples for which GEE converges, while QLS does not. It is also important to remember, that unlike GEE, the current implementation of QLS for the unstructured matrix requires a constant number of measurements per subject. However, as demonstrated for one example in Section 4.7, QLS can estimate the regression parameter more efficiently than GEE for a 3×3 unstructured matrix.

Next, we will replace the default estimator that is implemented for GEE in **xtgee** with Equation (5.25) that was obtained by Park (1993) as the maximum-likelihood estimator for Gaussian data when the standard deviations are not constrained to be equal, and that was also suggested by Chaganty and Shults (1999) as an alternate estimator should the stage two QLS estimate for the unstructured matrix fail to be positive definite: (This was obtained using the user-authored command **xtgeePark** (Shults, 2011).)

```
. xtgeePark change month month2, i(id) t(month) vce(robust) f(gau) c(uns)

  Results of GEE analysis:
```

GEE population-averaged model			Number of obs	=	48
Group and time vars:		id __00001C	Number of groups	=	8
Link:		identity	Obs per group: min =		6
Family:		Gaussian	avg =		6.0
Correlation:		fixed (specified)	max =		6

```
                                              Wald chi2(2)        =      41.74
Scale parameter:                  .7799752    Prob > chi2         =     0.0000


---------------------------------------------------------------------------
    change |    Coef.    Std. Err.       z     P>|z|     [95% Conf. Interval]
-----------+---------------------------------------------------------------
     month | -.0824896    .0139214    -5.93    0.000    -.1097752   -.0552041
    month2 |  .0017166    .0002749     6.25    0.000     .0011779    .0022554
     _cons |   1.32108    .2460223     5.37    0.000     .8388847    1.803274
---------------------------------------------------------------------------
Estimated correlation matrix:
```

The **xtgeePark** command will also automatically display the estimated correlation matrix and its eigenvalues.

```
Estimated within-id correlation matrix R:

         c1        c2        c3        c4        c5        c6
r1   1.0000
r2   0.7649    1.0000
r3   0.5663    0.9265    1.0000
r4   0.4488    0.7685    0.9229    1.0000
r5   0.5328    0.7960    0.9279    0.9674    1.0000
r6   0.9138    0.5560    0.3779    0.3639    0.4658    1.0000
Confirm that the eigenvalues of R are positive:

v[1,6]
          e1          e2          e3          e4          e5          e6
r1   4.4760292   1.1834237   .28310644   .0290255   .02288725   .00552787
```

The eigenvalues are all positive, although the sixth eigenvalue is close to zero. It is interesting that the estimated correlation between the fifth and sixth measurement occasions was ≈ 0.47 for Park's moment estimator, while it was ≈ 0.98 for QLS. The estimates and significance levels of the estimated regression coefficients also differed quite a bit between the two approaches. This suggests that it will be helpful to compare the fit of these two working structures.

The estimator (5.27) that we applied in this example could also be applied to unbalanced data; however, it is only guaranteed to be positive definite for complete data.

5.7 Other Structures

GEE and QLS use the same estimating equation for β and therefore yield identical results for an identity structure. Structures for analysis of familial data were discussed in Chapter 4, while structures for data with multiple sources of correlation will be presented in Chapter 6. In addition, although they have not yet been made available in the software packages that are currently available for implementation of QLS, several other working structures have been discussed in the literature for QLS. These include the banded Toeplitz for analysis of repeated bouts of measurements, in Shults et al. (2006a); an autoregressive familial correlation structure for analysis of balanced one-parent familial data with a structure that relates correlation to subject age, in Sabo and Chaganty (2010a); and a structure appropriate for a growth curve model, in Shi and Chaganty (2004).

5.8   Implementation of QLS for Patterned Correlation Structures

5.8.1   Algorithm for Implementation of QLS Using Software That Allows for Application of a User-Specified Working Correlation Structure That Is Treated as Fixed and Known in the GEE Estimating Equation for $\boldsymbol{\beta}$

Implementation of QLS is straightforward with a software package that allows for application of GEE with a fixed correlation structure; the updating step for estimation of $\boldsymbol{\beta}$ within each iteration of the QLS estimation procedure can then use the existing GEE software to solve the GEE estimating equation for $\boldsymbol{\beta}$, evaluated at the current QLS estimate of α. In the following algorithm, GEEPROC is a generic label for a software package that allows for implementation of GEE with a user-specified correlation structure; for example, it could refer to **xtgee** in Stata, or **geepack** in R.

1. Obtain a starting value for $\widehat{\boldsymbol{\beta}}$ by assuming $\alpha = 0$ and then implementing GEE with GEEPROC and a fixed identity working correlation structure.

2. **Stage One of QLS:** Alternate between the following steps until there is convergence in the estimates of $\boldsymbol{\beta}$:

   (a) Obtain updated values of the Pearson residuals at the current estimates of $\boldsymbol{\beta}$ and of α.

   (b) Update the estimate $\widehat{\alpha}_{QLSONE}$ of α by obtaining the solution to the stage one estimating equation (3.40) for α that is evaluated at the updated values of the Pearson residuals.

   (c) Construct the estimated working correlation structure $R(\widehat{\alpha}_{QLSONE})$ that corresponds to the updated estimate of α. For structures other than Markov, the matrix $R(\widehat{\alpha}_{QLSONE})$ will be constructed for the maximum value of n_i. For example, in a study in which the maximum number of observations per subject is 8 and the working correlation structure is AR(1), $R(\widehat{\alpha}_{QONE})$ will be an 8×8 AR(1) structure evaluated at $\widehat{\alpha}_{QONE}$. For the Markov structure, the dimension of $R(\widehat{\alpha}_{QONE})$ will equal the number of distinct values of the timing variable. For example, in a study in which some subjects are measured at times $\{1,2,8\}$ and all other subjects are measured at times $\{1,6,9\}$, the dimension of $R(\widehat{\alpha}_{QONE})$ will be 5×5.

   (d) Update the estimate of $\boldsymbol{\beta}$ by using the GEEPROC procedure to solve the GEE estimating equation (3.39) for $\boldsymbol{\beta}$, with a correlation structure that is treated as fixed and equal to $R(\widehat{\alpha}_{QONE})$.

3. **Stage Two:** After convergence in stage one, update the estimate of α by obtaining the solution to the stage two estimating equation (3.41) for α.

4. Construct the estimated working correlation structure $R(\widehat{\alpha}_{QLS})$ that corresponds to the stage two estimate of α.

5. Obtain the final estimate of $\boldsymbol{\beta}$ by using the GEEPROC procedure, with a correlation structure that is treated as fixed and equal to $R(\widehat{\alpha}_{QLS})$.

The **xtqls** command in Stata (Shults et al., 2007) and the **qlsinr** package in R (Xie and Shults, 2009) employ the above algorithm for implementation of QLS. One benefit of using available software for implementation of GEE within the estimation procedure for GEE is that post-estimation commands may be available for the GEE procedure. For example, **xtqls** (Shults et al., 2007) calls up the **xtgee** command for GEE. The **xtgee** command, which was developed by James Hardin for Stata, has many useful post-estimation commands, including **xtcorr** that we used earlier in this chapter to print out the estimated correlation matrix for QLS. However, there are some potential limitations. For example, the **qlsinr** package in R (Xie and Shults, 2009) calls up the **geepack** package (Yan, 2002; Halekoh et al., 2006). The **geepack** package is excellent but does not allow for application of a model-based covariance matrix; as result, one limitation of **qlsinr** is that it only allows for implementation of sandwich-based or jackknifed standard errors.

5.8.2   When Software for GEE Is Not Available, or Is Not Utilized

Ratcliffe and Shults (2008) developed the **QLSPACK** toolbox for MATLAB, allowing for the implementation of both GEE and QLS in MATLAB. (Prior to **QLSPACK** there was no toolbox available for implementation of GEE in MATLAB.) In addition, although SAS allows for implementation of GEE via the SAS GENMOD procedure, Kim and Shults (2010) directly programmed QLS without making use of PROC GENMOD, in the SAS macro %QLS. The **QLSPACK** toolbox and %QLS SAS macro both use an iterative approach for solution of the GEE estimating equation within each step of the estimation procedure; as described in Shults (1996) and Shults and Chaganty (1998), this approach makes use of the Cholesky decomposition of the inverse of the working correlation structure when solving the GEE estimating equation (3.39) for $\boldsymbol{\beta}$.

The algorithm implemented in **QLSPACK** for MATLAB and %QLS for SAS solves the GEE estimating equation within each stage of the QLS algorithm that is described in Section 3.4 as follows.

At the current estimate $\widehat{\alpha}$ of α (where $\widehat{\alpha} = \widehat{\alpha}_{QONE}$ in stage one versus $\widehat{\alpha} = \widehat{\alpha}_{QLS}$ in stage two), it first obtains the Cholesky decomposition of the inverse, $R_i^{-1}(\widehat{\alpha}) = L_i(\widehat{\alpha})L_i'(\widehat{\alpha})$.

Next, to solve the GEE estimating equation for $\boldsymbol{\beta}$ (evaluated at $\widehat{\alpha}$) it employs a modified Fisher-scoring method that updates the previous estimate by calculating the adjustment in Equation (2.22) as follows. First, it calculates $T_i = L_i'(\widehat{\alpha})\widetilde{A}_i^{-1}\left(Y_i - \widetilde{\boldsymbol{\mu}}_i\right)$ and $S_i = L_i'(\widehat{\alpha})\widetilde{A}_i^{-1}\widetilde{D}_i$, for $i = 1,\ldots,m$. Next, it regresses $T = (T_1',\ldots,T_m')'$ on $S = \left(S_1',\ldots,S_m'\right)'$, to obtain the adjustment $\left(\sum_{i=1}^{m}\widetilde{D}_i'\widetilde{V}_i^{-1}\widetilde{D}_i\right)^{-1}\left(\sum_{i=1}^{m}\widetilde{D}_i'\widetilde{V}_i^{-1}(Y_i - \widetilde{\boldsymbol{\mu}}_i)\right)$ that is added to the previous estimate of β in Equation (2.22). This iterative process continues until the adjustment is approximately zero.

5.9 Summary

QLS has several attractive features that were demonstrated in this chapter. It allows for straightforward implementation of some correlation structures that are not yet available in the major software packages that implement GEE; furthermore, it yields only one estimator for a particular working structure, while the choice of moment estimator is not always clear for the original formulation of GEE. QLS can also be applied as an alternative to GEE if the GEE estimation procedure should fail to converge; however, we caution the reader that any failure to converge should cause the analyst to carefully evaluate their choice of working correlation structure. QLS can also be easily implemented using software that is currently available for GEE, using the algorithm provided in Section 5.8. If no software for GEE is available, which was the case for MATLAB (Ratcliffe and Shults, 2008), then the method of QLS can be fully programmed using the approach discussed in Section 5.8.2.

With the exception of the unstructured matrix that was implemented using **xtqls2**, the examples in this chapter were obtained using the **xtqls** command in Stata (Shults et al., 2007). On the website for this book, we also provide commands to replicate many of the examples using the **QLSPACK** toolbox in MATLAB (Ratcliffe and Shults, 2008); the **qlsinr** package in R (Xie and Shults, 2009); and the QLS macro in SAS (Kim and Shults, 2010).

5.10 Exercises

Exercise 5.1 *Consider the exchangeable correlation structure (5.2). Show that the inverse of the exchangeable structure can be expressed as in (5.3). Also verify that the derivative of the inverse can be expressed as in Equation 5.4).*

Exercise 5.2 *For a given value of $\mathbf{Z}$, directly solve Equation (5.23) for the special case that $n = 2$.*

Exercise 5.3 *Show that the estimator you obtained in Exercise 5.2 is identical to the QLS stage one estimator for the AR(1) structure that was provided in Equation (3.31), when $n_i = 2 \; \forall \; i$ in Equation (3.31).*

Exercise 5.4 *Show that the estimator you obtained in Exercise 5.2 is identical to the QLS stage one estimator for the exchangeable structure that can be obtained by directly solving Equation (5.6), when $n_i = 2 \; \forall \; i$ in Equation (3.31).*

Exercise 5.5 *Show that the estimator you obtained in Exercise 5.2 is identical to the QLS stage one estimator for the tri-diagonal structure that can be obtained by directly solving Equation (5.11), when $n_i = 2 \; \forall \; i$ in Equation (3.31).*

Exercise 5.6 *Prove that* Z *defined in Equation (5.18) will be positive definite if* $n_i =$
$n \; \forall \; i$ *and* $m \geq n$.

Exercise 5.7 *For a* 3×3 *unstructured matrix show that the stage two QLS estimators
provided in Equations (4.18) and in (5.24) are identical.*

Exercise 5.8 *Show that when* $\delta_{ij} = 1$ *for all i, j in Equations (5.16) and (5.17), then
the solutions to these estimating equations are identical to the solutions provided for
the AR(1) structure in Equations (3.31) and (3.37), respectively.*

5.11   Appendix

Here we redo the analysis that we conducted in Section 4.5 with the working structure
for (mother, father, child) trios that was defined in structure (4.4) and that is identical
to a 3×3 unstructured correlation matrix. In the earlier example we obtained a stage
one estimate by directly solving the stage one estimating equations for the 3×3
structure and a stage two estimate using Equation (4.18). Here we redo the earlier
analysis: using the techniques described in this chapter for an $n \times n$ unstructured
matrix. The output and checks are not shown here. Only the final results, which
agree with those that we obtained earlier in Section 4.5, are shown.

```
xtqls2 bwt male first byr midage highage, i(family) t(type) /*
*/ c(uns) f(gau) vce(robust)
.

.

Updated matrix in stage two:

symmetric __000002[3,3]
             r1          r2          r3
c1    1
c2    .00771737         1
c3    .19660481   .23738574          1

Iteration 1: tolerance = .31949001
Iteration 2: tolerance = 6.090e-14
```

```
GEE population-averaged model              Number of obs      =      3000
Group and time vars:      family __00001C   Number of groups   =      1000
Link:                        identity       Obs per group: min =         3
Family:                      Gaussian                      avg =       3.0
Correlation:          fixed (specified)                    max =         3
                                           Wald chi2(5)       =    161.53
Scale parameter:              233090.7     Prob > chi2        =    0.0000

                           (Std. Err. adjusted for clustering on family)
-------------------------------------------------------------------------
             |              Semirobust
        bwt  |      Coef.    Std. Err.      z    P>|z|    [95% Conf. Interval]
-------------+-----------------------------------------------------------
       male  |   158.1303    17.91808     8.83   0.000    123.0115    193.2491
      first  |  -139.1568    18.83117    -7.39   0.000   -176.0652   -102.2484
        byr  |   3.639571    .6847018     5.32   0.000    2.29758    4.981562
```

```
   midage |    56.44828    30.63661      1.84    0.065    -3.598368    116.4949
  highage |    118.4659    59.92465      1.98    0.048     1.015732     235.916
    _cons |      3218.2    57.54763     55.92    0.000     3105.409    3330.992
---------------------------------------------------------------------------
```

Chapter 6

Analysis of Data with Multiple Sources of Correlation

6.1 Characteristics of Data with Multiple Sources of Correlation

The typical GEE or QLS analysis involves data with one level (or source) of correlation within the clusters. For example, in a longitudinal study it is often reasonable to assume that measurements between subjects are independent, so that the only source of correlation is due to the similarity between the repeated measurements on a subject. However, if the longitudinal study simultaneously assessed systolic and diastolic blood pressure on patients, then we might anticipate two sources of correlation, due to the fact that measurements within a patient might tend to be more similar if they represent the same type of blood pressure, or are measured more closely together in time. Analyses such as this that assess multiple longitudinal outcomes simultaneously are also referred to as *multivariate longitudinal* analyses.

Studies with three or more sources of correlation are also possible. For example, if both systolic and diastolic blood pressure were measured simultaneously in a longitudinal study of spouses, then three sources of correlation might be anticipated, due to the fact that measurements within a family might be more similar if they are measured on the same spouse, represent the same type of blood pressure, or are collected on the same measurement occasion.

In this chapter we describe the implementation of QLS for analysis of data with multiple sources of correlation. We describe several types of multi-source correlated data that are totally balanced, balanced within clusters, or are unbalanced. We describe how to model the association of multi-source correlated data and how to implement the appropriate correlation structures for analysis with QLS. As we present our results, we will introduce new notation as needed.

6.2 Multi-Source Correlated Data That Are Totally Balanced

6.2.1 Example of Multivariate Longitudinal Data That Are Totally Balanced

In this chapter we consider data with two sources of correlation, from the study described in Section 1.5.4 of Chapter 1. In this study, three methods of suctioning a patient's breathing tube were compared at five measurement occasions. The suctioning data might be expected to have two sources of correlation, due to the fact that two measurements on a subject might tend to be more similar if they are collected more closely together in time, or if they represent the same type of suctioning method. These data were *totally balanced* because all subjects had five measurements collected for each of three suctioning methods; the total number of measurements for each patient was fifteen.

Although the notation and methods in this chapter will focus on data with two sources of correlation, that is, on *multivariate longitudinal data*, the extension to data with three or more sources of correlation is simple. For example, Shults and Ratcliffe (2009) consider data with three sources of correlation. Usually, at most two or three sources of correlation will be considered in an analysis, although four or more sources of correlation may occur.

6.2.2 *Notation*

The usual notation for GEE is easily extended for data with multiple sources of correlation. For example, in the suctioning study, if oxygen saturation was only measured on one method, then there would only be one source of correlation in the data, due to the fact that measurements on a subject might tend to be more similar if they were measured more closely together in time. In this situation, the measurement and associated covariates collected at time j on subject i would be represented by y_{ij} and x_{ij} respectively, which involve subscript i and one additional subscript that represents the one source of correlation in this study. To extend this notation to data with two sources of correlation, we simply increase the total number of subscripts to three, so that $y_{ij_1j_2}$ and $x_{ij_1j_2}$ represent the value of the outcome variable and associated $p \times 1$ vector of covariates that are collected on subject i when the values of the first and second sources of correlation are j_1 and j_2, respectively. In general, all the notation for GEE and QLS can be extended in a similar fashion, by simply increasing the number of subscripts so that there is one subscript (i) that indicates the subject (or cluster) and w additional subscripts ($j_1, j_2, \cdots j_w$) that represent the w additional sources of correlation in the data. The data are totally balanced when $j_1 = 1, \cdots, n_1$ through $j_w = 1, \cdots, n_w$; the total number of measurements per subject is then $n_1 \times n_2 \times \cdots n_w$, which is the same for all subjects.

In the suctioning study, n_1 represents the number of methods of suctioning and equals three for all patients, while n_2 represents the number of measurement occasions per subject and equals five for all patients.

Next, it will be helpful to refer to $Y_i[a, b]$ as the vector of outcomes of measurements $y_{ij_1j_2}$ on subject i that has been sorted first according to j_a, and then with respect to j_b. For example, in the suctioning study $Y_i[1, 2]$ represents the vector of measurements on patient i that has been first sorted according to subscript j_1 (that represents method of suctioning) and then with respect to j_2 (that represents timing of measurement):

$$Y_i[1,2] = (y_{i11}, y_{i12}, y_{i13}, y_{i14}, y_{i15}, y_{i21}, y_{i22}, y_{i23}, y_{i24}, y_{i25}, y_{i31}, y_{i32}, y_{i33}, y_{i34}, y_{i35})'.$$

Next, $Y_i[2, 1]$ represents the vector of measurements on patient i that has been first sorted according to subscript j_2 (that represents timing of measurement) and then with respect to j_1 (that represents method of suctioning):

$$Y_i[2,1] = (y_{i11}, y_{i21}, y_{i31}, y_{i12}, y_{i22}, y_{i32}, y_{i13}, y_{i23}, y_{i33}, y_{i14}, y_{i24}, y_{i34}, y_{i15}, y_{i25}, y_{i35})'.$$

As mentioned in Section 2.3.4 for data with one source of correlation, throughout this text we distinguish between the working structure $R_i(\alpha)$ and the true true structure $T_i(\rho)$. As in earlier chapters, we assume that the true structure has been correctly specified, unless specified otherwise. In this chapter we will begin by constructing working correlation structures $R[2,1]$ for outcome vectors such as $Y_i[2,1]$, using the structures that were defined in Section 2.3.3. (Similarly, we could use $T[2,1]$ to denote the true structures that depend on the true structures defined in Section 2.3.4.) We then consider data that are balanced within subjects, followed by data that are unbalanced. Throughout, we use boldface font to indicate the working correlation structures that we show can be constructed using the structures provided in Section 2.3.4. (Similarly, we will use boldface font to denote the true structures.)

6.2.3 Working Correlation Structure for Balanced Data

There is a straightforward approach that can be used to specify a reasonable working correlation structure for data with multiple sources of correlation. Here we continue with the suctioning example that has two sources of correlation; however, the approach is easily extended for data with more than two sources, for example, the following description was also provided in Shults and Ratcliffe (2009), but for data with three sources of correlation.

1. First, within each cluster (or subject), label each source of correlation, so that one source will be the first source (and will be represented by subscript j_1 in $y_{ij_1j_2}$) while the other source will be the second (and will be represented by subscript j_2 in $y_{ij_1j_2}$). The order of sources does not matter, as the labeling is only done to facilitate organizing and sorting the data.

2. For each source of correlation, identify a working correlation structure that would be appropriate if that particular source was the only source of correlation in the data. Let $R_a(\alpha_a)$ represent the working structure for source a; $a = 1, 2$.

3. After identifying a working structure for each source of correlation, construct the correlation structure for the vector of measurements $Y_i[1,2]$ as the Kronecker product of $R_1(\alpha_1)$ and $R_2(\alpha_2)$. The working correlation structure of $Y_i[1,2]$ is then given by $R_i[1,2] = R[1,2] = R_1(\alpha_1) \otimes R_2(\alpha_2)$. It is also easy to show that $R_i(a,b) = R_a(\alpha_a) \otimes R_b(\alpha_b)$, where (a,b) is any permutation of $(1,2)$, that is, $(a,b) \in \{(1,2),(2,1)\}$. (Note that $R_i[1,2] = R[1,2]$; that is, the working structure does not depend on subscript i because the working structure is the same for all subjects, for balanced data.)

For example, in the suctioning study we labeled type of suctioning method as the first source of correlation, while timing of measurement was the second source.

To identify an appropriate working structure for the first source of correlation (type of method), we simply pretend that this is the only source of correlation in the data, that is, that each patient had measurements collected with each of the three types of suctioning methods, on one measurement occasion. If we do not have any reason to believe that oxygen saturation levels will tend to be more similar for any two of the three suctioning methods, a reasonable structure $R_1(\alpha_1)$ for type of method might be the 3×3 exchangeable structure that was described in Section 2.3.3 of Chapter 2:

$$
R_1(\alpha_1) = \begin{pmatrix} 1 & \alpha_1 & \alpha_1 \\ \alpha_1 & 1 & \alpha_1 \\ \alpha_1 & \alpha_1 & 1 \end{pmatrix}.
$$

Next, to specify a plausible working structure for the second source of correlation (timing of measurement), we pretend that the only source of correlation is timing of measurement, so that each patient has five measurements that were collected via one particular method of suctioning. Assuming that the measurements are approximately equally spaced in time, a reasonable structure $R_2(\alpha_2)$ to describe the pattern of association due to timing is the AR(1) that was also described in Section 2.3.3 of Chapter 2:

$$
R_2(\alpha_2) = \begin{pmatrix}
1 & \alpha_2 & \alpha_2^2 & \alpha_2^3 & \alpha_2^4 \\
\alpha_2 & 1 & \alpha_2 & \alpha_2^2 & \alpha_2^3 \\
\alpha_2^2 & \alpha_2 & 1 & \alpha_2 & \alpha_2^2 \\
\alpha_2^3 & \alpha_2^2 & \alpha_2 & 1 & \alpha_2 \\
\alpha_2^4 & \alpha_2^3 & \alpha_2^2 & \alpha_2 & 1
\end{pmatrix}.
$$

The working correlation structure for $\boldsymbol{Y}_i[1,2]$ is then constructed by taking the Kronecker product of $R_1(\alpha_1)$ and $R_2(\alpha_2)$, $\boldsymbol{R}[1,2] = R_1(\alpha_1) \otimes R_2(\alpha_2)$, which can be expressed as a partitioned matrix:

$$
\begin{pmatrix}
R_2(\alpha_2) & \alpha_1 R_2(\alpha_2) & \alpha_1 R_2(\alpha_2) \\
\alpha_1 R_2(\alpha_2) & R_2(\alpha_2) & \alpha_1 R_2(\alpha_2) \\
\alpha_1 R_2(\alpha_2) & \alpha_1 R_2(\alpha_2) & R_2(\alpha_2)
\end{pmatrix}. \tag{6.1}
$$

In the above partitioned matrix, the matrices $R_2(\alpha_2)$ on the diagonal represent the (working) within-patient pattern of association for one method of suctioning over time. The other matrices $\alpha_1 R_2(\alpha_2)$ represent the (working) within-patient pattern of association between two methods of suctioning over time. In this representation, the parameter α_1 dampens the correlation between two measurements if they were collected via two methods of suctioning. For example, according to this Kronecker product structure, the correlation between two measurements on a subject that were collected at visits 1 and 4 will be α_2^3 if they were collected with the same type of suctioning method, but will be $\alpha_1 \alpha_2^3$ if they were collected with two different methods; the latter correlation will be smaller in absolute value than the former, which is plausible because they have a greater degree of disagreement with respect to the sources of correlation in the data due to their representing two different methods of suctioning.

In general, if $R[1,2] = R_1(\alpha_1) \otimes R_2(\alpha_2)$, then using the definition of Kronecker product, the assumed correlation between $y_{ij_1j_2}$ and $y_{ij'_1j'_2} = R_1[j_1,j'_1] \times R_2[j_2,j'_2]$, where $R_1[j_1,j'_1]$ is element (j_1,j'_1) of matrix $R_1(\alpha_1)$. If two measurements *agree* with respect to a source of correlation, say the first, then their correlation can be expressed as $R_1[j_1,j_1] \times R_2[j_2,j'_2]$, which equals $R_2[j_2,j'_2]$ because the diagonal elements $R_1[j_1,j_1]$ of correlation matrix $R_1(\alpha_1)$ equal one. However, if the measurements *disagree* with respect to the first source of correlation, then their correlation is $R_1[j_1,k_1] \times R_2[j_2,j'_2]$, which is less than or equal to $R_2[j_2,j'_2]$ in absolute value. As a result, the correlation between any two measurements will be smaller in absolute value if they have a greater degree of disagreement with respect to the sources of correlation in the data. The Kronecker product structure is therefore an attractive structure for the analysis of multi-source correlated data, because it is easy to set up and it has the property that the correlation between measurements is related to their degree of agreement with respect to the sources of correlation in the data.

6.2.4 Prior Implementation of the Kronecker Product Structure

One of the earliest authors to propose an approach for a GEE analysis of multivariate longitudinal data was Dr. Myrto Lefkopoulou of Harvard University, who unfortunately died at the age of 34 after battling cancer for 2 years. Lefkopoulou et al. (1989) considered binary data with two sources of correlation. They assumed a working correlation structure that can be expressed as a Kronecker product structure for data that are balanced within clusters. They also made two simplifying assumptions. First, they assumed that the covariates do not vary within clusters (e.g., using response time as a covariate for the suction example). They then reduced their data to a series of cluster-level proportions, which resulted in an overdispersed binomial model with a correlation structure that is equivalent to the usual correlation structure for data with one source of correlation. They then implemented GEE under an assumption of a constant scalar parameter for all clusters.

Galecki (1994) described implementation of the Kronecker product structure for analysis of data with two or more repeated factors (sources of correlation). Galecki (1994) was cited in the documentation for the **MIXED** procedure for **SAS**, which starting with version 6.12 allows for implementation of a Kronecker product structure for analysis of normally distributed data with mixed linear models. Roy and Khattree have also done quite a bit of work on the implementation of Kronecker product structures in analysis of normally distributed data (Roy and Khattree, 2005, 2003). Other authors, for example, Lu and Zimmerman (2005) and Naik and Rao (2001), have implemented Kronecker products for analysis of normal data. Shults and Morrow (2002) and Shults and Ratcliffe (2009) implemented a Kronecker product structure for QLS analysis of data that are balanced within clusters, while Chaganty and Naik (2002) implemented this structure for totally balanced data. Kim and Shults (2010) and Kim et al. (2008b) then implemented a Kronecker product structure for unbalanced data.

6.2.5   Implementation of QLS for Analysis

In the following section we will describe an algorithm that can be used to implement the Kronecker product structure for analysis of data with multiple sources of correlation that are either totally balanced or that are balanced within clusters. Here we demonstrate the Stata command **xtmultcorr2** (Shults and Ratcliffe, 2009) for implementation of QLS for analysis of data with two sources of correlation. See Shults and Ratcliffe (2009) for more detailed instructions on how to implement the **xtmultcorr** Stata commands.

We will use the dataset that was described in Section 1.5.4 of Chapter 1. The outcome variable was oxygen saturation, which was measured on each patient at baseline and then at each of four subsequent measurement occasions.

We first open the dataset in Stata and describe the variables:

```
. use example_multilevel, clear
. describe

Contains data from example_multilevel.dta
  obs:            375
  vars:             6                            30 Nov 2006 09:06
  size:        19,500 (98.1% of memory free)
-------------------------------------------------------------------------------
> --
             storage  display     value
variable name  type    format     label       variable label
-------------------------------------------------------------------------------
> --
id              double  %10.0g                 subject id
time            double  %10.0g                 measurement occasion
type            double  %10.0g                 method of suctioning
o2              double  %10.0g                 oxygen saturation
family          double  %10.0g                 artificial family variable
high            double  %10.0g                 1 if o2>96; 0 otherwise
-------------------------------------------------------------------------------

Sorted by:  id  type  time
```

Next we check the number of subjects:

```
. *Number of subjects?
. qui tab id

. noi di _result(2)
25
```

There are 25 subjects. Next we cross-tabulate *type* and *time*, to see if the data are balanced overall:

```
. tab type time
```

method of suctioning	measurement occasion					Total
	1	2	3	4	5	
1	25	25	25	25	25	125
2	25	25	25	25	25	125

```
        3 |        25          25          25          25         25 |       125
-----------+----------------------------------------------------------+----------
   Total |        75          75          75          75         75 |       375
```

The above tabulation shows that the data are balanced overall, because there are 25 measurements for each combination of method and measurement occasion. We also note that **xtmultcorr2** will check for overall balance, and will return an error if the data are not totally balanced.

Next, we generate indicator variables *method1*, *method2*, and *method3* that each take value one (and zero otherwise) if the method of suctioning is type 1, 2, or 3, respectively.

```
. *Generate an indicator variable for  type of method
. qui tab type, gen(method)
```

Next, we fit a model that regresses oxygen saturation on *time* and method of suctioning. We include the two indicator variables *method2* and *method3*, so that the reference group is method 1. In our analysis we adjust for the two sources of correlation described in Section 6.2.3. (Although *type* and *time* are labeled as the first and second sources of correlation, respectively, the order could have been reversed, so that *time* is first and *type* is second.) We specify the equicorrelated structure for source 1 (*type*) and the AR(1) structure for source 2 (*time*). We also specify the sandwich covariance matrix for estimation of the covariance matrix of $\widehat{\beta}$:

```
. xtmultcorr2 o2 time method2 method3, i(id) l1(type) l2(time)/*
*/  c1(exc) c2(AR 1)  f(gau) vce(robust) f(gau) vce(model)
Estimated correlation associated with level one:

symmetric __000003[3,3]
          c1          c2          c3
r1         1
r2  .15541311           1
r3  .15541311   .15541311           1

Estimated correlation associated with level two:

symmetric __000004[5,5]
          c1          c2          c3          c4          c5
r1         1
r2  .67744482           1
r3  .45893149   .67744482           1
r4  .31090076   .45893149   .67744482           1
r5  .21061811   .31090076   .45893149   .67744482           1

Iteration 1: tolerance = .04715547
Iteration 2: tolerance = 3.186e-14

GEE population-averaged model                   Number of obs      =        375
Group and time vars:            id __00001T      Number of groups   =         25
Link:                              identity      Obs per group: min =         15
Family:                            Gaussian                     avg =       15.0
Correlation:             fixed (specified)                      max =         15
                                                Wald chi2(3)       =      10.89
Scale parameter:                   7.484291     Prob > chi2        =     0.0123
```

```
                                   (Std. Err. adjusted for clustering on id)
-----------------------------------------------------------------------------
             |             Semirobust
         o2  |     Coef.   Std. Err.      z    P>|z|     [95% Conf. Interval]
-------------+---------------------------------------------------------------
        time |   .0855847   .0354605    2.41   0.016     .0160834    .155086
     method2 |   .7843529   .4311426    1.82   0.069    -.060671    1.629377
     method3 |   .4178262   .4562029    0.92   0.360    -.476315    1.311967
       _cons |   95.25977   .4546379  209.53   0.000     94.3687    96.15084
-----------------------------------------------------------------------------
```

What we see from the above output is that the estimated correlation due to source 1 (type of method of suctioning) is $\widehat{\alpha}_1 \approx 0.1554$, which is smaller than the estimated correlation due to source 2 (order of measurement within subjects), $\widehat{\alpha}_2 \approx 0.6774$. We also see that there appears to be some suggestion that oxygen saturation levels are rising over time and that the second method of suctioning might have slightly higher levels than method 1. (However, we do note that the goal of this analysis is merely to demonstrate the method.)

Next, we fit the same model for the binary outcome *high*, that takes value one if $o_2 > 96$ and takes value 0 otherwise.

```
. xtmultcorr2 high time method2 method3, i(id) l1(type) /*
*/ l2(time)  c1(exc)  c2(AR 1)  f(bin 1) vce(robust)

Estimated correlation associated with level one:

symmetric __000003[3,3]
           c1          c2          c3
r1          1
r2  .16879769           1
r3  .16879769   .16879769           1

Estimated correlation associated with level two:

symmetric __000004[5,5]
           c1          c2          c3          c4          c5
r1          1
r2  .64954332           1
r3  .42190652   .64954332           1
r4  .27404656   .42190652   .64954332           1
r5  .17800511   .27404656   .42190652   .64954332           1

Iteration 1: tolerance = .16302989
Iteration 2: tolerance = .00079797
Iteration 3: tolerance = 6.696e-06
Iteration 4: tolerance = 1.397e-07

GEE population-averaged model                Number of obs      =        375
Group and time vars:            id __00001T   Number of groups   =         25
Link:                                 logit   Obs per group: min =         15
Family:                            binomial                  avg =       15.0
Correlation:              fixed (specified)                  max =         15
                                             Wald chi2(3)       =       3.20
Scale parameter:                          1   Prob > chi2       =     0.3613
```

```
                               (Std. Err. adjusted for clustering on id)
      ------------------------------------------------------------------------
              |             Semirobust
       high   |    Coef.    Std. Err.      z    P>|z|     [95% Conf. Interval]
      --------+---------------------------------------------------------------
       time   |  .0302963   .0511407     0.59   0.554    -.0699375    .1305302
    method2   |  .4728172   .3676909     1.29   0.198    -.2478436    1.193478
    method3   |  .2470832    .349464     0.71   0.480    -.4378536    .9320201
      _cons   |  -.702204   .3951834    -1.78   0.076    -1.476749    .0723413
      ------------------------------------------------------------------------
```

The estimated correlations were similar for the continuous and binary outcomes; for both, the estimated correlation within subjects over *time* (source 2 of correlation) was greater than the correlation between methods of suctioning (source 1).

Before moving onto the next section we attempt to fit an unstructured correlation matrix, because this might provide insight into the pattern of association within clusters.

We start by sorting on *id*, *type*, and *time*, and then creating a variable *order* that takes value 1,..., 15 and indicates the order of measurements within each cluster.

```
sort id type time
by id: gen order = _n
```

Next, we attempt to replicate the example with the continuous outcome *o2*. The following command indicates that the variable *order* is the timing variable for this example.

```
. xtgee o2 time method2 method3, i(id) c(uns) robust t(order)

Iteration 1: tolerance = .12644357
.

.
Iteration 11: tolerance = 7.684e-07

GEE population-averaged model                Number of obs     =        375
Group and time vars:              id order    Number of groups  =         25
Link:                             identity    Obs per group: min =        15
Family:                           Gaussian                   avg =       15.0
Correlation:                  unstructured                   max =         15
                                             Wald chi2(3)      =       9.14
Scale parameter:                  7.544736    Prob > chi2       =     0.0275

                               (Std. Err. adjusted for clustering on id)
      ------------------------------------------------------------------------
              |             Semirobust
        o2    |    Coef.    Std. Err.      z    P>|z|     [95% Conf. Interval]
      --------+---------------------------------------------------------------
       time   | -.0367815   .0574352    -0.64   0.522    -.1493523    .0757894
    method2   |  1.020284   .3498434     2.92   0.004     .3346034    1.705964
    method3   |  .5779569   .4187593     1.38   0.168    -.2427962     1.39871
      _cons   |   95.2619   .4419627   215.54   0.000     94.39567    96.12813
      ------------------------------------------------------------------------
convergence not achieved
r(430);
```

As indicated above, convergence was not achieved for the unstructured matrix.

We next attempt to replicate the example with a binary outcome, but with an unstructured matrix.

```
. xtgee high time method2 method3, i(id) f(bin 1) c(uns) robust t(order)

Iteration 1: tolerance = .48569069
.

.
Iteration 22: tolerance = 6.591e-07
```

```
GEE population-averaged model              Number of obs      =        375
Group and time vars:            id order   Number of groups   =         25
Link:                              logit   Obs per group: min =         15
Family:                         binomial                  avg =       15.0
Correlation:               unstructured                  max =         15
                                           Wald chi2(3)       =       9.02
Scale parameter:                       1   Prob > chi2        =     0.0291

                                 (Std. Err. adjusted for clustering on id)
-------------------------------------------------------------------------
             |              Semirobust
        high |    Coef.    Std. Err.      z    P>|z|    [95% Conf. Interval]
-------------+-----------------------------------------------------------
        time |   .020247    .0455327    0.44   0.657   -.0689954    .1094894
     method2 |  .5322152    .3132279    1.70   0.089   -.0817002    1.146131
     method3 | -.2777616     .232252   -1.20   0.232   -.7329671    .1774438
       _cons | -.7515678    .2810368   -2.67   0.007    -1.30239   -.2007458
-------------------------------------------------------------------------
convergence not achieved
r(430);
```

The above output shows that after 22 iterations, convergence was not achieved.

To assess the difficulty with respect to convergence, we next obtain the estimated working correlation structure and its eigenvalues. (The matrix is 15×15 and so is not printed.)

```
. matrix R = e(R)
. matrix symeigen x eig = R
. matrix list eig

eig[1,15]
           e1          e2          e3          e4          e5          e6
r1  5.4963958   3.5125044   2.0431563    1.121775   1.0400217   .54036354

           e7          e8          e9         e10         e11         e12
r1  .31599335   .29986394   .25467981   .20862141   .16948214   .11775427

          e13         e14         e15
r1  .04408824  -.02996527  -.13473463
```

The above output indicates that some of the eigenvalues are negative, so that the estimated unstructured matrix was not positive definite; this resulted in a failure to converge in the GEE estimation procedure. The 15×15 unstructured matrix involves 105 correlation parameters. In general, multi-source correlated data will tend to involve larger clusters than analyses with one source of correlation. As a result, there could be problems with respect to application of an unstructured matrix for other analyzes of multi-source data, if the cluster sizes become too large and the GEE estimation procedure fails to converge (as it did for this example).

6.3   Multi-Source Correlated Data That Are Balanced within Clusters

6.3.1   Example

It is not difficult to modify the design of the suctioning study in Section 6.2, for which the data are totally balanced, so that the study data are balanced only *within* subjects (or clusters). Suppose that in the suctioning study, some patients did not have their final measurement, and that another subgroup of patients were only assessed with two methods of suctioning; in this situation, the data would not be totally balanced, but would be balanced *within* patients. In general for this study, the data will be balanced within patients when the same methods of suctioning are applied at every visit (e.g., if all methods are applied at baseline on a subject, then they are all applied at all subsequent visits for that subject) and the same assessments are available for each method (e.g., if two measurements were taken via method one for a particular patient then two measurements will be available for the other methods as well).

6.3.2   Notation

It is easy to extend the notation for totally balanced multi-source correlated data to data that are balanced within subjects (or clusters). The notation presented in Section 6.2.2 for totally balanced data will remain the same, except that the subscripts will not necessarily take the same values for all subjects. For data with two sources of correlation that are balanced within subjects, the subscript j_1 that corresponds to the first source of correlation will take value $j_1 = 1, \cdots, n_{i1}$, while subscript j_2 that corresponds to the second source of correlation will take value $j_2 = 1, \cdots, n_{i2}$.

For example, suppose that subject 3 is missing all measurements at the final measurement occasion. Then her vector of measurements will be represented by

$$\boldsymbol{Y}_3[1,2] = (y_{i11}, y_{i12}, y_{i13}, y_{i14}, y_{i21}, y_{i22}, y_{i23}, y_{i24}, y_{i31}, y_{i32}, y_{i33}, y_{i34})'.$$

For subject 3, the number of methods of assessment (at each visit), n_{31}, equals 3, while the number of measurement occasions (for each method of assessment), n_{32}, equals 4.

Or, if subject 2 only had measurements collected during visit 1 with methods of suctioning 1 and 3, then her vector of measurements would be represented by

$$\boldsymbol{Y}_2[1,2] = (y_{i11}, y_{i21})'.$$

For this subject, the number of methods of assessment (at each visit) n_{21} equals 2 and the number of visits (for each method of assessment) n_{22} equals 1. Note also that because we assume an exchangeable structure for source 1 (type of correlation), it does not matter how we label the methods within each visit, that is, in this example we labeled methods 1 and 3 with $j_1 = 1$ and $j_1 = 2$, respectively.

6.3.3   Correlation Structure for Data That Are Balanced within Clusters

To set up correlation structures for data that are balanced within clusters, we implement the same algorithm that was described in Section 6.2.3, the only difference being that when we identify a correlation structure for each source of correlation in the data, the dimensions of the structure will be allowed to vary between subjects.

For example, in the suctioning study, the data were totally balanced and the specified working correlation structure was $R[1,2] = R_1(\alpha_1) \otimes R_2(\alpha_2)$, where $R_1(\alpha_1)$ was a 3×3 exchangeable correlation structure, and $R_2(\alpha_2)$ was a 5×5 AR(1) structure.

To extend this for data that are balanced within subjects, we would set up a working correlation structure for subject i as $R_i[1,2] = R_{i1}(\alpha_1) \otimes R_{i2}(\alpha_2)$, where $R_{i1}(\alpha_1)$ is an $n_{i1} \times n_{i1}$ exchangeable correlation structure and $R_{i2}(\alpha_2)$ is an $n_{i2} \times n_{i2}$ AR(1) structure.

For example, consider the subjects discussed in Section 6.3.2. The working correlation structure for patient 3 (who is missing all measurements at the final measurement occasion) would be $R_{31}(\alpha_1) \otimes R_{32}(\alpha_2)$, where $R_{31}(\alpha_1)$ is a 3×3 exchangeable correlation structure and $R_{32}(\alpha_2)$ is a 4×4 AR(1) structure. For subject 3 the dimension of the AR(1) structure is only 4 because she is missing measurements at the final visit.

In addition, the working correlation structure for patient 2 (who only has two methods of assessment measured only at visit 1) would be $R_{21}(\alpha_1) \otimes R_{22}(\alpha_2)$, where $R_{21}(\alpha_1)$ is a 2×2 exchangeable correlation structure, and $R_{22}(\alpha_2)$ is a 1×1 AR(1) structure. A 1×1 correlation matrix is equal to 1, so that the working correlation structure for patient 2 is $R_{21}(\alpha_1) \otimes 1 = R_{21}(\alpha_1)$.

6.3.4   Algorithm for Implementation of QLS for Multi-Source Correlated Data That Are Balanced within Clusters

Sorting and premultiplying will be important for implementing QLS for data with multiple sources of correlation. In fact, we can implement QLS for data with multiple sources of correlation using QLS estimators for data with *one source of correlation* if we sort and premultiply appropriately. Next, we explain why this is the case. In this section we assume that the true structure has been correctly specified, so that the true and working structures are identical. In other words, here we assume that $Corr(Y_i[a,b]) = T_i[a,b] = R_i[a,b]$, where $T_i[a,b] = T_{ia}(\rho_a) \otimes T_{ib}(\rho_b)$ and $R_i[a,b] = R_{ia}(\alpha_a) \otimes R_{ib}(\alpha_b)$.

Suppose that we first sort our data according to source a and then source b, for $(a,b) \in \{(1,2),(2,1)\}$. We then define the Pearson residuals by $Z_i[a,b] = A_i^{-1/2}(a,b)(Y_i[a,b] - \mu_i[a,b])$, where $A_i(a,b) = diag(h(u_{i11}), \ldots, h(u_{in_an_b}))$.

Next we premultiply the residuals $Z_i[a,b]$ by $R_{ia}^{-1/2}(\alpha_a) \otimes I_{n_{ib}}$, where $R_{ia}^{-1/2}(\alpha_a)$ is the square root of the inverse of $R_{ia}(\alpha_a)$ to obtain

$$Z_i^*[a,b] = \left(R_{ia}^{-1/2}(\alpha_a) \otimes I_{n_{ib}}\right) Z_i[a,b].$$

It is then easy to show that

$$Corr\left(\mathbf{Z}_i^*[a,b]\right) = I_{n_{ia}} \bigotimes R_{ib}(\alpha_b). \tag{6.2}$$

Structure (6.2) is the correlation structure of n_{ia} uncorrelated subjects, each of whom has correlation structure $R_{ib}(\alpha_b)$. Sorting and premultiplying therefore yielded data with one source of correlation.

In what follows we define the estimated Pearson residuals as $\widehat{\mathbf{Z}}_i[a,b] = \widehat{A}_i^{-1/2}(a,b)\left(\mathbf{Y}_i[a,b] - \widehat{\boldsymbol{\mu}}_i[a,b]\right)$. Our algorithm for implementation of QLS within *stage one* of the estimation procedure therefore alternates between sorting first with respect to source 1 and then with respect to source 2; premultiplying $\widehat{\mathbf{Z}}_i[1,2]$ by $\left(R_{i1}^{-1/2}(\widehat{\alpha}_1) \bigotimes I_{n_{i2}}\right)$; and then updating the estimate of α_2 using formulas obtained earlier for data with one source of correlation (for structure $R_i(\alpha_2)$). After updating the estimate for α_2, we next sort the data with respect to source 2 and then with respect to source 1; premultiply $\widehat{\mathbf{Z}}_i[2,1]$ by $\left(R_{i2}^{-1/2}(\widehat{\alpha}_2) \bigotimes I_{n_{i1}}\right)$; and then update the estimate of α_1 using formulas obtained earlier for data with one source of correlation (for structure $R_i(\alpha_1)$). (Because this description applies to both stage one and two of the estimation procedure, we used the generic $\widehat{\alpha}_a$ to represent the current estimate of α_a.) We will see later in this section that for data that are totally balanced (not just balanced *within* clusters), we will have a similar simplification for stage two of QLS, so that the usual equations for data with one source of correlation can be applied in stage two, for data that are totally balanced within subjects.

Next we provide a more formal algorithm for this process, for data with two sources of correlation. (See Shults et al. (2004) for a general algorithm for data with multiple sources of correlation.) In what follows, the estimated $\widehat{\mathbf{Z}}_i^*[a,b] = \left(R_{ia}^{-1/2}(\widehat{\alpha}_a) \bigotimes I_{n_{ib}}\right)\widehat{\mathbf{Z}}_i[a,b]$, for $(a,b) \in \{(1,2),(2,1)\}$.

Algorithm for Implementation of QLS for Data that are Balanced Within Clusters

1. Obtain a starting value for $\widehat{\boldsymbol{\beta}}$ by assuming $\alpha_1 = \alpha_2 = 0$ (so that the working correlation structures are identity matrices) and then solving the GEE estimating equation (3.39) for $\boldsymbol{\beta}$.

2. **Stage One of QLS:** Alternate between the following steps until there is convergence in the estimates of $\boldsymbol{\beta}$:

   (a) *Updating Step for Correlation Parameters:* Alternate between Step $[1,2]$ and Step $[2,1]$ until there is convergence in the estimates of α_w (for $w = 1,2$), where Step $[a,b]$ is defined as follows:

   Step $[a,b]$: Sort the data according to source a and then source b. Obtain the premultiplied Pearson residuals $\widehat{\mathbf{Z}}_i^*[a,b]$ that are evaluated at the current estimate of α_a. Next, obtain an updated estimate of α_b as the solution (for α_b) to the QLS stage one estimating equation

$$\sum_{i=1}^{m}\sum_{j=1}^{n_{ia}} \widehat{\mathbf{z}}_{ij}^{*\prime}[a,b]\frac{\delta R_{ib}^{-1}(\alpha_b)}{\delta \alpha_b}\widehat{\mathbf{z}}_{ij}^{*}[a,b] = 0, \tag{6.3}$$

where $\widehat{Z}_{ij}^*[a,b]$ is the vector that consists of rows $(j-1)n_b$ to jn_b of vector $\widehat{Z}_i^*[a,b]$.

(b) *Updating Step for the Regression Parameter:* After updating the estimates of α_1 and α_2, construct the estimated working correlation structure $R_{i1}(\widehat{\alpha}_{QONE1}) \otimes R_{i2}(\widehat{\alpha}_{QONE2})$ that is evaluated at the current estimates of α_1 and of α_2.

(c) Update the estimate of $\boldsymbol{\beta}$ by solving the GEE estimating equation (3.39) evaluated at the working correlation structures $R_{i1}(\widehat{\alpha}_{QONE1}) \otimes R_{i2}(\widehat{\alpha}_{QONE2})$, for $\boldsymbol{\beta}$. The software for implementation of this algorithm in **Stata** that is described in Shults and Ratcliffe (2009) uses the **xtgee** command to solve Equation (3.39) with the working structure treated as fixed and known, and equal to $R_{i1}(\widehat{\alpha}_{QONE1})) \otimes R_{i2}(\widehat{\alpha}_{QONE2})$. The software for **Matlab** that is described in Shults and Ratcliffe (2009) solves Equation (3.39) using the approach described in Section 5.8.2.

3. **Stage Two:** After convergence in stage one, update the estimates of α_1 and of α_2 by obtaining the solution $(\widehat{\alpha}_{QLS1}, \widehat{\alpha}_{QLS2})$ to the stage two estimating equations for (α_1, α_2):

$$\sum_{i=1}^m trace\left(\left.\frac{\partial R_{ia}^{-1}(\alpha_a)}{\partial \alpha_a}\right|_{\alpha_a=\widehat{\alpha}_{QONEa}} R_{ia}(\alpha_a)\right) f_i(b) = 0, \qquad (6.4)$$

where $f_i(b) = trace\left(\left.R_{ib}^{-1}(\alpha_b)\right|_{\alpha_b=\widehat{\alpha}_{QONEb}} R_{ib}(\alpha_b)\right)$; $(a,b) \in \{(1,2),(2,1)\}$, and $\widehat{\alpha}_{QONEw}$ is the QLS stage one estimate of α_w, for $w = 1,2$. If the data are totally balanced, then $f_i(b)$ is a constant and the stage two estimating equations reduce to the usual stage two estimating equations for data with one source of correlation.

4. Construct the estimate of working structure $R_{i1}(\alpha_1) \otimes R_{i2}(\alpha_2)$ that is evaluated at the stage two estimates $(\widehat{\alpha}_{QLS1}, \widehat{\alpha}_{QLS2})$ of α_1 and α_2.

5. Obtain the final estimate of $\boldsymbol{\beta}$ by solving the GEE estimating equation (3.39) evaluated at the working correlation structures $R_{i1}(\widehat{\alpha}_{QLS1}) \otimes R_{i2}(\widehat{\alpha}_{QLS2})$.

The estimating equations (6.3) and (6.4) are very similar to the stage one and stage two estimating equations (3.40) and (3.41), respectively. Given that we can implement QLS for multi-source correlated data by sorting, premultiplying, and using estimators obtained earlier for data with one source of correlation in stage one of the procedure (and in stage two as well, for totally balanced data), it is not difficult to program QLS for data with multiple sources of correlation. As mentioned earlier, software is available in Stata and MATLAB that implements the algorithm that is described above, in Shults and Ratcliffe (2009).

6.3.5   *Implementation of QLS for Analysis*

Next we continue with the example of Section 6.2.5. We start by dropping some observations, so that as in Section 6.3.2, patient 3 has no measurements at the final visit, while patient 2 only has measurements at visit 1, that were taken with two methods of assessment.

```
. drop if time==5 & id==3
(3 observations deleted)

. drop if time>1 & id==2
(12 observations deleted)

. drop if type==3 & id==2
(1 observation deleted)
```

Next, note that if we cross-tabulate *type* and *time* within groups of subjects who share the same values for n_{i1} and n_{i2}, then the data will be balanced within the groups. First, if we cross-tabulate within subjects who have $n_{i1} = 3$ and $n_{i2} = 4$ (which is only patient 3) we see that the number of measurements per combination of *type* and *time* is identical.

```
. tab type time if id==3
```

method of suctioning	measurement occasion 1	2	3	4	Total
1	1	1	1	1	4
2	1	1	1	1	4
3	1	1	1	1	4
Total	3	3	3	3	12

We observe the same balance if we cross-tabulate within subjects for whom $n_{i1} = 2$ and $n_{i2} = 1$ (which is only patient 2).

```
. tab type time if id==2
```

method of suctioning	measurement occasion 1	Total
1	1	1
2	1	1
Total	2	2

Next, if we cross-tabulate *type* and *time* for the third group of subjects, for whom $n_{i1} = 3$ and $n_{i2} = 5$, then we observe that the data are balanced within this group as well.

```
. tab type time if id<2 | id>3
```

method of suctioning	measurement occasion 1	2	3	4	5	Total
1	23	23	23	23	23	115

```
         2 |        23          23          23          23          23 |        115
         3 |        23          23          23          23          23 |        115
-----------+-----------------------------------------------------------+----------
     Total |        69          69          69          69          69 |        345
```

In general it will be difficult to check whether the data are balanced within subjects by cross-tabulating, especially when there are more than two sources of correlation in the data. The **xtmultcorr** commands will do the checking for us. For example, next we attempt to fit the last model that we fit prior to deleting observations, using **xtmultcorr2** for balanced data.

```
. xtmultcorr2 high time method2 method3, i(id) l1(type) /*
*/ l2(time)  c1(exc)  c2(AR 1)  f(bin 1) vce(robust)

Data are not balanced overall with respect to sources of
correlation.
```

We were given a warning that the data are not balanced overall, that is, they are not totally balanced with respect to the sources of correlation in the data. We therefore next fit the same model using **xtmultcorr2u** that is appropriate for data that are balanced within subjects.

```
. xtmultcorr2u high time method2 method3, i(id) l1(type) /*
*/ l2(time)  c1(exc)  c2(AR 1)  f(bin 1) vce(robust)
Estimated correlation associated with level one:

symmetric __000003[3,3]
            c1          c2          c3
r1           1
r2   .17571777           1
r3   .17571777   .17571777            1

Estimated correlation associated with level two:

symmetric __000004[5,5]
            c1          c2          c3          c4          c5
r1           1
r2   .64663724           1
r3   .41813972   .64663724           1
r4   .27038472   .41813972   .64663724           1
r5   .17484083   .27038472   .41813972   .64663724           1

Iteration 1: tolerance = .12929074
Iteration 2: tolerance = .00236245
Iteration 3: tolerance = .00003729
Iteration 4: tolerance = 5.307e-07

GEE population-averaged model              Number of obs       =        359
Group and time vars:           id __00001W  Number of groups    =         25
Link:                                logit  Obs per group: min =          2
Family:                            binomial                 avg =       14.4
Correlation:              fixed (specified)                 max =         15
                                            Wald chi2(3)        =       3.97
Scale parameter:                         1  Prob > chi2         =     0.2644

                                 (Std. Err. adjusted for clustering on id)
```

```
             |              Semirobust
        high |      Coef.    Std. Err.       z    P>|z|     [95% Conf. Interval]
-------------+---------------------------------------------------------------
        time |    .0448386    .0523282     0.86   0.392    -.0577228    .1473999
     method2 |    .4417547    .3463119     1.28   0.202    -.2370041    1.120514
     method3 |    .1668288    .3338016     0.50   0.617    -.4874104    .8210679
       _cons |   -.6782476    .3909927    -1.73   0.083    -1.444579    .0880841
-------------------------------------------------------------------------------
```

The above results are almost identical to those that were obtained for balanced data at the end of Section 6.2.5, which is reasonable because the datasets were almost identical for the two analyses. However, note that in the output above, the observations per group range from a minimum of two to a maximum of fifteen. When the data were totally balanced, the number of measurements was the same for all subjects.

We also note that the more general command **xtmultcorr2u** could have been applied in Section 6.2.5, because data that are totally balanced are also balanced within subjects. However, for totally balanced data, we strongly recommend the application of **xtmultcorr2**, because it is much faster than **xtmultcorr2u**. In addition, commands **xtmultcorr3** (and **xtmultcorr3u**) are available for analysis of totally balanced (and balanced within subject) data with three sources of correlation; these programs are described and demonstrated in Shults and Ratcliffe (2009). These programs can be slow!

6.4   Multi-Source Correlated Data That Are Unbalanced

Here we describe multi-source correlated data that are unbalanced. We implement the approach developed by Kim et al. (2008b).

6.4.1   Example

In Sections 6.2 and 6.3 we described data that are totally balanced and that are balanced *within* subjects, respectively. Here we consider a particular type of *unbalanced* data, for which the observed data are a subset of totally balanced measurements.

For example, suppose that in the suctioning study the investigators planned to use three methods of suctioning at each of five measurement occasions. If all measurements were taken as planned, then the data would be totally balanced, as in Sections 6.2. If some measurements were missing, but in such a way that the same methods of suctioning were always utilized for a particular subject, then the data would be balanced *within* subjects, as in Section 6.3. However, if measurements were missing so that the number of methods of suctioning was not the same at each visit for a particular subject, then the data would be *unbalanced*.

Data from a clinical trial will often be balanced within subjects, because missing data are often due to missed visits. For example, in a trial that measures visual acuity on the left and right eyes of each subject, we might anticipate that as long as the subject shows up for her visit, then acuity will be measured on both of her eyes. In this situation, the data will be balanced within subjects because the number of visits may vary between subjects, but within subjects the same number of measurements (two) are taken at each visit. However, it is possible to obtain unbalanced data. For example, a subject who has only one measurement taken on the first visit due to a left eye infection, but who has two measurements taken on subsequent visits, will yield unbalanced data. In this situation, the number of measurements per visit is not constant within subjects.

6.4.2   Notation

For unbalanced data, we use the notation of Section 6.2. However, we also need to define matrix $E_i(1,2)$, which is the matrix that is obtained by removing the rows of an identity matrix corresponding to the missing measurements for subject i. Another way to view $E_i(1,2)$ is as the matrix that *retains* the rows that correspond to the observed measurements for subject i, when the data for subject i have first been sorted with respect to source 1 and then with respect to source 2. (Matrix $E_i(2,1)$ pertains to data for subject i that have been sorted first with respect to source 2 and then with respect to source 1.)

For example, in the suctioning study, n_1 (the planned number of methods of suctioning) and n_2 (the planned number of visits per subject) satisfy $n_1 = 3$ and $n_2 = 5$. The working correlation structure for subjects with complete measurements therefore has dimension $n_1 \times n_2 = 3 \times 5 = 15$.

If we label *type* as the first source of correlation and label *time* as the second source and sort the data according to *type* and *time*, the indices for measurements on subject i will be

$$\{11, 12, 13, 14, 15, 21, 22, 23, 24, 25, 31, 32, 33, 34, 35\}.$$

Suppose that the second subject has measurements taken with method 1 at visits 1 and 2; with method 2 at visits 1 and 5; and no measurements taken with method 3. This subject therefore has measurements with the following indices:

$$\{11, 12, 21, 25\}.$$

Matrix $E_2(1,2)$ is therefore defined by *retaining* the 1st, 2nd, 6th, and 10th rows of a 15×15 identity matrix:

$$E_2(1,2) = \begin{pmatrix} 1 & 0 & 0 & 0 & 0 & 0 & 0 & 0 & 0 & 0 & 0 & 0 & 0 & 0 & 0 \\ 0 & 1 & 0 & 0 & 0 & 0 & 0 & 0 & 0 & 0 & 0 & 0 & 0 & 0 & 0 \\ 0 & 0 & 0 & 0 & 0 & 1 & 0 & 0 & 0 & 0 & 0 & 0 & 0 & 0 & 0 \\ 0 & 0 & 0 & 0 & 0 & 0 & 0 & 0 & 0 & 1 & 0 & 0 & 0 & 0 & 0 \end{pmatrix}. \qquad (6.5)$$

6.4.3   Correlation Structure for Data That Are Unbalanced

We consider unbalanced data that comprise a subset of measurements from a study that planned for balanced measurements. Suppose we consider the vectors of measurements on each subject that have first been sorted according to source 1 and then source 2. We can then define the working correlation structure for each vector of measurements by first defining the $n_1 n_2 \times n_1 n_2$ structure $R_1(\alpha_1) \otimes R_2(\alpha_2)$ for a subject with complete data (the *complete* structure). Then, the working structure for subject i will be defined by pre and post-multiplying with $E_i(1,2)$ and $E_i(1,2)'$, respectively, where $E_i(1,2)$ is the $n_1 n_2 \times n_1 n_2$ identity matrix with rows removed that correspond to the missing observations on subject i. Pre- and post- multiplication with $E_i(1,2)$ and $E_i(1,2)'$ selects the rows and columns of the complete structure that correspond to the observed measurements for subject i. The correlation structure for subject i is then defined as $R_i[1,2] = E_i(1,2)(R_1(\alpha_1) \otimes R_2(\alpha_2)) E_i(1,2)'$.

For example, consider subject 2 who was described in Section 6.4.2, for whom matrix $E_2(1,2)$ was defined in Equation (6.5). The working structure for this subject will be defined as $R_2(1,2) = E_2(1,2)(R_1(\alpha_1) \otimes R_2(\alpha_2)) E_2(1,2)'$, where $R_1(\alpha_1) \otimes R_2(\alpha_2)$ was defined in Equation (6.1). Subject 2 had measurements taken at the 1st, 2nd, 6th and 10th measurement occasions. Pre- and post-multiplication by $E_2(1,2)$ and $E_2(1,2)'$ selects the 1st, 2nd, 6th and 10th rows of $R_1(\alpha_1) \otimes R_2(\alpha_2)$. The working correlation structure for subject 2 can also be simplified as follows:

$$R_2[1,2] = \begin{pmatrix} 1 & \alpha_2 & \alpha_1 & \alpha_1 \alpha_2^4 \\ \alpha_2 & 1 & \alpha_1 \alpha_2 & \alpha_1 \alpha_2^3 \\ \alpha_1 & \alpha_1 \alpha_2 & 1 & \alpha_2^4 \\ \alpha_1 \alpha_2^4 & \alpha_1 \alpha_2^3 & \alpha_2^4 & 1 \end{pmatrix}. \tag{6.6}$$

Matrix (6.6) represents the correlation matrix of measurements with indices $\{11, 12, 21, 25\}$. Recall that according to our Kronecker product working structure, the assumed correlation between y_{ijk} and $y_{ij'k'}$ is $R_1[j,j']R_2[k,k']$; this could also have been used to quickly define the elements of the working structure for subject 2. For example, the assumed correlation between y_{211} and $y_{225} = R_1[1,2]R_2[1,5] = \alpha_1 \alpha_2^4$.

Note that if $R_1(\alpha_1)$ and $R_2(\alpha_2)$ are positive definite, then it is easy to show that $R_1(\alpha_1) \otimes R_2(\alpha_2)$ is positive definite. In addition, it is easy to show that $R_i[1,2] = E_i(1,2)\{R_1(\alpha_1) \otimes R_2(\alpha_2)\} E_i(1,2)'$ will be positive definite, as follows. Let x be a non-zero vector. Then, $x'R_i[1,2]x = x'E_i(1,2)\{R_1(\alpha_1) \otimes R_2(\alpha_2)\} E_i(1,2)'x = y'\{R_1(\alpha_1) \otimes R_2(\alpha_2)\}y$, where it is easy to show that $y \neq 0$ because $x \neq 0$. Because $R_1(\alpha_1) \otimes R_2(\alpha_2)$ is positive definite, it follows that

$$y'\{R_1(\alpha_1) \otimes R_2(\alpha_2)\}y > 0.$$

Therefore, $R_i[1,2] = E_i(1,2)\{R_1(\alpha_1) \otimes R_2(\alpha_2)\} E_i(1,2)'$ is positive definite.

6.4.4   Algorithm for Implementation of QLS for Multi-Source Correlated Data That Are Unbalanced

Implementation of QLS for unbalanced data is more complex because the correlation structure of unbalanced data no longer has the Kronecker product structure, but instead is a sub-matrix of a Kronecker product of matrices. As a result, the properties of Kronecker product structures no longer hold. For example, the inverse of the correlation matrix can no longer be easily obtained as the Kronecker product of inverses that was the inverse of a Kronecker product structure. As a result, the following algorithm for unbalanced data involves estimating equations that are more complex than those presented earlier for data that are balanced within subjects, in Section 6.3.4. In addition, it involves the working structure $R_i[a,b]$ that was defined in Section 6.4.3. It also utilizes the expression for the first derivative of the inverse of a matrix that was defined in Equation (5.10).

Algorithm for Implementation of QLS for Data that are Unbalanced

1. Obtain a starting value for $\widehat{\boldsymbol{\beta}}$ by assuming $\alpha_1 = \alpha_2 = 0$ (so that the working correlation structures are identity matrices) and then solving the GEE estimating equation (3.39) for $\boldsymbol{\beta}$.

2. **Stage One of QLS:** Sort the data according to source 1 and then source 2. Alternate between the following steps until there is convergence in the estimates of $\boldsymbol{\beta}$:

   (a) *Updating Step for Correlation Parameters:* Obtain the updated Pearson residuals $\widehat{\mathbf{Z}}_i[1,2]$ that are evaluated at the current estimate of $\boldsymbol{\beta}$. Next, alternate between Steps [1] and [2] until there is convergence in the estimates of α_w (for $w = 1,2$):

   Step [1]: Obtain an updated estimate $\widehat{\alpha}_{QONE1}$ of α_1 as the solution (for α_1) to the QLS stage one estimating equation

   $$\sum_{i=1}^{m} \widehat{\mathbf{Z}}_i[1,2]' \boldsymbol{R}_i^{-1}[1,2] \left[E_i(1,2) \left\{ \frac{\delta R_1(\alpha_1)}{\delta \alpha_1} \otimes R_2(\widehat{\alpha}_{QONE2}) \right\} E_i(1,2)' \right]$$
   $$\times \boldsymbol{R}_i^{-1}[1,2] \widehat{\mathbf{Z}}_i[1,2] = 0, \qquad (6.7)$$

   where $E_i(1,2)$ was defined in Section 6.4.2 and $\boldsymbol{R}_i^{-1}[1,2]$ is evaluated at $(\alpha_1, \widehat{\alpha}_{QONE2})$.

   Step [2]: Obtain an updated estimate $\widehat{\alpha}_{QONE2}$ of α_2 as the solution (for α_2) to the QLS stage one estimating equation

   $$\sum_{i=1}^{m} \widehat{\mathbf{Z}}_i[1,2]' \boldsymbol{R}_i^{-1}[1,2] \left[E_i(1,2) \left\{ R_1(\widehat{\alpha}_{QONE1}) \otimes \frac{\delta R_2(\alpha_2)}{\delta \alpha_2} \right\} E_i(1,2)' \right]$$
   $$\times \boldsymbol{R}_i^{-1}[1,2] \widehat{\mathbf{Z}}_i[1,2] = 0, \qquad (6.8)$$

   where $\boldsymbol{R}_i^{-1}(1,2)$ is evaluated at $(\widehat{\alpha}_{QONE1}, \alpha_2)$

   (b) *Updating Step for the Regression Parameter:* After updating the estimates of α_1 and α_2, construct the estimated working correlation structures $\widehat{\boldsymbol{R}}_i[1,2]$ $= E_i(1,2) \left\{ R_1(\widehat{\alpha}_{QONE1}) \otimes R_2(\widehat{\alpha}_{QONE2}) \right\} E_i(1,2)'$.

   (c) Update the estimate of $\boldsymbol{\beta}$ by solving the GEE estimating equation (3.39) evaluated at the current estimates $\widehat{\boldsymbol{R}}_i[1,2]$ of working correlation structures $\boldsymbol{R}_i[1,2]$. The software for implementation of this algorithm Kim et al. (2008b) uses the approach described in Section 5.8.2 to solve the GEE estimating equation for $\boldsymbol{\beta}$.

3. **Stage Two:** After convergence in stage one, update the estimates of α_1 and of α_2 by obtaining the solutions $\widehat{\alpha}_{QLSa}$ (for α_a) to the stage two estimating equations for $a = 1, 2$:

$$\sum_{i=1}^{m} trace\left(\left\{\boldsymbol{R}_i^{-1}[1,2]\frac{\delta\boldsymbol{R}_i[1,2]}{\delta\alpha_a}\boldsymbol{R}_i^{-1}[1,2]\right\}E_i(1,2)\{R_1(\alpha_1)\otimes R_2(\alpha_2)\}E_i(1,2)'\right) = 0,$$

(6.9)

where $\boldsymbol{R}_i[1,2]$ and $\boldsymbol{R}_i^{-1}[1,2]$ are evaluated at the stage one estimates $(\widehat{\alpha}_{QONE1}, \widehat{\alpha}_{QONE2})$ of (α_1, α_2).

4. After updating the estimates of α_1 and α_2, construct the estimated working correlation structures $\widehat{\boldsymbol{R}_i[1,2]} = E_i\{R_1(\widehat{\alpha}_{QLS1})\otimes R_2(\widehat{\alpha}_{QLS2})\}E_i'$.

5. Obtain the final estimate of $\boldsymbol{\beta}$ by solving the GEE estimating equation (3.39) evaluated at estimated working correlation structures $\widehat{\boldsymbol{R}}_i = E_i\{R_1(\widehat{\alpha}_{QLS1})\otimes R_2(\widehat{\alpha}_{QLS2})\}E_i'$.

6.4.5 Implementation of QLS for Analysis

There is currently no software available in Stata for analysis of multivariate longitudinal data. However, version 2 of the SAS macro % QLS allows for analysis of data with multiple sources of correlation (Kim and Shults, 2013). We implement this software here.

Before moving to SAS, within Stata we will randomly drop some observations so that our data will be unbalanced. First, we open up the dataset:

```
. clear
. set seed 99999
. use example_multlevel, clear
```

Next, we generate a variable *temp* that is uniformly distributed over $(0,1)$. We drop all observations for which *temp* is less than 0.20, which will cause us to randomly drop approximately 20 percent of the measurements in the dataset. Note that we set the seed for the random number generator to 99999, so that all readers should obtain the same results for this example.

```
. set seed 99999
. gen temp = uniform()
. drop if temp < .2
(74 observations deleted)
. tab type time

. tab type time
```

method of suctioning	measurement occasion					Total
	1	2	3	4	5	
1	16	20	22	22	18	98
2	20	21	22	21	22	106
3	19	19	21	19	19	97
Total	55	60	65	62	59	301

In the cross-tabulation of *time* and *type* above, we see that the data are clearly not totally balanced. Next we fit the **xtmultcorr2u** command in Stata, which will run without error if the data are *balanced within subjects.*

```
. xtmultcorr2u high time method2 method3, i(id) l1(type) /*
*/ l2(time) c1(exc) c2(AR 1)  f(bin 1) vce(robust)
Data are not balanced within subjects with respect to sources of
correlation.
```

We received a warning message that the data are not balanced within subjects. Next we outsheet a dataset that will be a text file that we will then read into SAS:

```
outsheet id time type method2 method3 family high /*
*/ using sas_example_multi.txt, comma replace
```

Next we go into SAS and run the macro %QLSV2, with the following commands: (Note that the directory will need to be changed below.)

```
data oxygen;
infile "h:\QLSbook\chapter5\sas_example_multi2.txt"
       delimiter = "," ;
input id time type method2 method3 family high ;
run;
%QLS(data=oxygen, y=high, x=method2 method3 time, time=time,
                   id=id, outcome=type, link=2, corr=5);
```

We obtain the following output, which is similar (but not identical) to the results obtained in Section 6.3.5 prior to dropping 20 percent of the measurements:

```
Analysis of Stage 2 QLS Parameter Estimates
```

Parameter	Estimate	Stand Err	Z	Pr>|Z|	[95% Con. Interval]
INTERCEPT	-0.668401	0.4902828	-1.36	0.1728	-1.629338 0.2925353
METHOD2	0.4884267	0.3513878	1.39	0.1645	-0.200281 1.177134
METHOD3	0.1866659	0.4098249	0.46	0.6488	-0.616576 0.989908
TIME	0.0217923	0.0855827	0.25	0.7990	-0.145947 0.1895313

```
        Rho       Alpha

     0.150563  0.6688907
```

The estimates Rho and Alpha given above are $\hat{\alpha}_{QLS1}$ and $\hat{\alpha}_{QLS2}$, respectively.

6.5 Asymptotic Relative Efficiency Calculations

Suppose that the true correlation structure for the suctioning study is the Kronecker product structure defined in (6.1). If no software is available for analysis of multivariate longitudinal data, a simpler approach would be to implement QLS (or GEE) with the exchangeable correlation structure (2.11) that is available in the major software packages that allow for implementation of QLS (or GEE). However, a drawback to this simpler approach is that it can result in a loss in efficiency in estimation of the regression parameter. Here we assess the loss of efficiency that would occur for the suctioning study if the true model for expected oxygen saturation was the estimated model for the continuous outcome (*o2*) in Section 6.2.5, and the true correlation structure was the Kronecker product structure (6.1) that was misspecified as exchangeable (2.11), so that we incorrectly assumed that the correlation between any two measurements on a subject was constant.

We assume that the regression parameter $\beta = (\beta_0, \beta_1, \beta_2, \beta_3)' = (95.2598, 0.0856, 0.7844, 0.4178)'$, where β_0 is the constant, and β_1, β_2, β_3 are the regression coefficients for *time*, *method2*, and *method3*, respectively. We assumed several different values for the correlation parameters, with the anticipation that we would suffer a greater loss in efficiency when the difference between the true and working structures was greatest.

To calculate the efficiencies, recall that for both QLS and GEE, $\sqrt{m}(\hat{\beta} - \beta)$ is asymptotically Gaussian with mean 0 and covariance matrix V_β given by (2.23). For a canonical link function, V_β can be simplified as follows:

$$V_\beta = \lim_{m \to \infty} \phi S_m \left\{ \sum_{i=1}^{m} X_i' A_i^{1/2} R_i^{-1} T_i R_i^{-1} A_i^{1/2} X_i \right\} S_m, \qquad (6.10)$$

where

$$S_m = \phi \left\{ \sum_{i=1}^{m} X_i' A_i^{1/2} R_i^{-1} A_i^{1/2} X_i \right\}^{-1}, \qquad (6.11)$$

and T_i and R_i are the true and working structures for subject i, respectively. If the working correlation structures were correctly specified, so that all $R_i = T_i$, then the covariance matrix V_β reduces to $V_T = \lim_{m \to \infty} S_M$.

We then compute the efficiency for $\hat{\beta}_j$ by dividing the asymptotic variance under correct specification by the asymptotic variance under incorrect specification, i.e. as the ratio of the jth diagonal element of V_β and the jth diagonal element of V_T. However, when we compute the asymptotic covariance matrices, we need to recall that the limiting value of the QLS estimate of the association parameter will not necessarily be consistent if the true structure is misspecified. We therefore need to calculate the efficiencies at the limiting value of the estimator of the association parameter for the exchangeable structure, as was suggested in Sutradhar and Das (1999). An algorithm to obtain the limiting value when the true Kronecker product is misspecified as exchangeable is provided in the appendix for this chapter. The efficiencies are calculated as the number of subjects $m \to \infty$; we assumed that the covariance design was replicated as subjects enrolled in the study.

Table 6.1 displays the efficiencies for QLS. The first line displays the results when α_1 and α_2 both equal zero, so that the true structure is an identity matrix. An identity structure is a special case of an exchangeable structure, so that the working exchangeable structure is correctly specified. As anticipated, the efficiencies are all one in this situation. Line 2 assesses the loss in efficiency if α_1 and α_2 were equal to their estimated values in the analysis of oxygen saturation. This indicates that we can suffer a serious loss in efficiency in this situation; for example, the efficiency for *time* is only 58 percent. Line 3 displays the situation when α_1 and α_2 are equal and small in value. As might be anticipated, the loss in efficiency is relatively small in this situation. However, lines 4 and 5 display the situation in which the correlations are larger in value. They demonstrate that the loss in efficiency can be severe. For example, the efficiency for *method2* is only 52 percent for $\alpha_1 = 0.40$ and $\alpha_2 = 0.60$.

The results shown in Table 6.1 indicate that incorrect application of the exchangeable structure can result in a substantial loss in efficiency in estimation of $\boldsymbol{\beta}$. We also note that the limiting value for the GEE moment estimator of the association parameter when the true Kronecker product structure is misspecified as exchangeable, is almost identical to the limiting value for the QLS estimator. The results shown here are therefore almost identical for GEE and QLS. We could therefore suffer an appreciable loss in efficiency if the true structure is a Kronecker product, but we fit an exchangeable structure with either QLS or GEE.

Table 6.1 *Percent Efficiencies for the Regression Coefficients for the Constant Term, Time, Method2, and Method3, When the True Kronecker Correlation Structure is Misspecified as Exchangeable. The true structure* $= R_1(\alpha_1) \otimes R_1(\alpha_1)$ *where* $R_1(\alpha_1)$ *is* 3×3 *exchangeable and* $R_2(\alpha_2)$ *is* 5×5 *AR(1). The working structure is* 15×15 *exchangeable.* **limit** *is the limiting value of the QLS estimator of the association parameter when the true Kronecker product structure was misspecified as exchangeable in the analysis of the suctioning study.*

α_1	α_2	limit	constant	time	method2	method3
1.00	1.00	1.00	1.00	1.00	1.00	1.00
0.16	0.68	0.21	0.79	0.58	0.61	0.76
0.20	0.20	0.07	0.96	0.90	0.90	0.97
0.30	0.40	0.14	0.91	0.76	0.73	0.92
0.40	0.60	0.26	0.85	0.57	0.52	0.83

6.6   Summary

The typical application of QLS, or GEE, is to data with one source of correlation. For example, if three monthly measurements are made on the left eye of each of thirty patients, there will be one source of correlation, due to the similarity between the repeated measurements within a patient. If monthly measurements are made on *both eyes* of the subjects, then there will be an additional source of correlation, due to the similarity between the left and right eyes of each patient. Additional sources of correlation are also possible. For example, if the measurements are made on brother–sister pairs, then there will be three sources of correlation, due to similarity within one eye over time, within the left and right eyes, and within the brother–sister pairs.

The Kronecker product structure is a plausible working correlation structure for the analysis of multi-source correlation data, because it forces the correlation between measurements to decline with increasing *disagreement* with respect to the sources of correlation in the data. For example, for the Kronecker product structure, the correlation between the left and right eye of a subject will be greater if they were both measured at visit 1, than if one eye was measured at visit 1, while the other was measured at visit 2.

The Kronecker product structure has nice features that have made it popular for the analysis of multi-source data. It is easy to construct for different settings, as the product of the structures that would be plausible for each source, if it were the only source of correlation in the data. For example, in the two-source ocular study described above, we could specify the Kronecker product of a 2 × 2 correlation structure (to represent correlation within the left and right eye at each visit) and a 3 × 3 AR(1) structure (to represent correlation within the repeated measurements on one eye over time).

The inverse of the Kronecker product of two matrices is simply the Kronecker product of the inverse of each matrix; this facilitates implementation of this structure for QLS or GEE analysis of data that are *completely balanced*, or that are *balanced within subjects*. In the ocular study, the data would be completely balanced if three measurements were made on the left and right eye of all participants. As a result, every subject would have the same number of measurements (six). The data would be balanced within subjects if, within a patient, the same number of measurements are made on each eye; in addition, within each patient, the same eyes are measured at each visit. For example, some patients could have one measurement on both eyes, while others could have three on both eyes, or two on the left eye. For data that are balanced within subjects, the total number of measurements can vary between subjects; however, the number of measurements on the left and right eyes must agree, and the same eyes must be measured at each visit.

Due to its ease of implementation for balanced data, most researchers have implemented the Kronecker product structure for totally balanced data. We demonstrated software in Stata for data that are completely balanced, and that are balanced within subjects.

The situation becomes more complicated when the data are unbalanced. Unbalanced data would result from the ocular study if some subjects had three measurements on the left eye and two on the right eye data would also be unbalanced if some patients had only their left eye measured at visit 1, and their right eye measured at visit 3. If the Kronecker product structure was the true structure for subjects with complete data, the structure for subjects with missing measurements would be obtained by deleting rows and columns from the full Kronecker product structure; the rows and columns to be removed correspond to the missing measurements. For example, a patient who is missing the first measurement would have a working structure that is obtained by removing the first row and column of the full structure.

Once rows and columns have been removed from the Kronecker product structure, the resulting structure can no longer be expressed as a Kronecker product, and the nice properties of the Kronecker product are lost. As a result, very few people have worked on implementation of a Kronecker product structure for unbalanced data. We described and demonstrated software that is available for analysis in Stata of data that are balanced within patients, or that are totally balanced. We also described and demonstrated software in SAS (Kim and Shults, 2013) that can be applied to unbalanced data. We also demonstrated that we can suffer a loss in asymptotic relative efficiency if we incorrectly ignore a source of correlation in our data, so that it is worthwhile to correctly model the correlation in multi-source correlated data with a true Kronecker product structure.

6.7 Exercises

Exercise 6.1 *Show that the result in Equation (6.2) is correct, so that*

$$Corr\left(\mathbf{Z}_i^*[a,b]\right) = I_{n_{ia}} \bigotimes R_{ib}(\alpha_b).$$

Exercise 6.2 *If the data are totally balanced, show that the estimating equations in (6.4) reduce to the usual estimating equations for data with one source of correlation.*

Exercise 6.3 *Propose plausible Kronecker product working correlation structures for the following study designs:*

   a. *Systolic and diastolic blood pressure will be measured on each of 20 patients at baseline, and at 6 and 12 months post baseline.*

   b. *Systolic and diastolic blood pressure will be measured on each member of 20 brother–sister pairs, at baseline, and at 6 and 12 months post baseline.*

   c. *Lead levels in the blood will be measured once annually for 5 years, on each person who resides within each of 90 geographic areas that are defined according to zip code. The number of people who live within each zip code varies between 1,000 and 5,000.*

Exercise 6.4 *Consider each of the study designs that were described in Exercise 6.3; do you anticipate that they will yield totally balanced data, data that are balanced within subjects, or data that are unbalanced?*

Exercise 6.5 *Consider the study in which the weights of each puppy in litters of three puppies is measured at birth, and then at 1 and 2 weeks of age. Describe why a Kronecker product structure might be reasonable to model the pattern of association among the weights of puppies within each litter. Describe a plausible structure for each of the following litters. (Hint: Set up the structure for litter with complete data, and then obtain the structures for litters with missing data by removing the rows and columns that correspond to the missing measurements.)*

   a. *A litter with complete data.*

   b. *A litter whose members were not measured at week 1, but were weighed at birth and at 2 weeks of age.*

   c. *A litter with all members measured at week 1, but with a missing weight on only one puppy at visit 2.*

6.8 Appendix: The Limiting Value of the QLS Estimate of the Association Parameter When the True Correlation Structure Is Misspecified as Exchangeable (from Appendix in Xie et al. (2010))

Here we assume that the true structure is a Kronecker product structure $T_1(\rho_1) \otimes R_2(\rho_2)$, while the working structure is $n_i \times n_i$ exchangeable $R_i(\alpha)$. In this section we therefore use R_i and α to represent the exchangeable working structure and its parameter, in order to distinguish between the true and working structures. Also, n_i represents the dimension of the true and working structures. The dimensions of $T_1(\rho_1)$ and $T_2(\rho_2)$ are assumed to be n_1 and n_2, respectively, so that $n_i = n_1 n_2$ for the efficiency calculations in this section.

The approach provided here is quite general and could be quickly adapted for any true structure, when the working structure is exchangeable. For example, a very similar approach was provided in Xie et al. (2010) to obtain the limiting value of $\hat{\alpha}_{QLS}$ when true mixed familial structures were misspecified as exchangeable. To modify the approach described here for a different true structure, simply substitute $T_1(\rho_1) \otimes T_2(\rho_2)$ with the alternate true structure and (ρ_1, ρ_2) with the alternate true correlation parameters in what follows.

Next, to obtain the limiting value of $\hat{\alpha}_{QLS}$ when the true Kronecker product structures are misspecified as exchangeable, we follow the approach provided in Theorem 3.2 of Chaganty and Shults (1999). First note that $E(Z_i(\beta) Z_i'(\beta)) = \phi T_1(\rho_1) \otimes T_2(\rho_2)$. It then follows that the solution to the stage one estimating equation (3.40) converges in probability to the solution (for α) to the following estimating equation:

$$trace\left(\sum_{i=1}^{m} \frac{\partial}{\partial \alpha} R_i^{-1}(\alpha) \left(T_1(\rho_1) \otimes T_2(\rho_2) \right) \right) = 0. \tag{6.12}$$

The inverse of exchangeable structure $R_i(\alpha)$ can be expressed as $R_i^{-1}(\alpha) = \frac{1}{(1-\alpha)} I_{n_i} - \frac{\alpha}{(1-\alpha)(1+(n_i-1)\alpha)} e_i e_i'$, where I_{n_i} is the identity matrix and e_i is an $n_i \times 1$ column vector of ones. Next, because $trace(e_i e_i'(T_1(\rho_1) \otimes T_2(\rho_2))) = trace(e_i'(T_1(\rho_1) \otimes T_2(\rho_2)) e_i) = e_i'(T_1(\rho_1) \otimes T_2(\rho_2)) e_i$, Equation (6.12) can easily be simplified as follows:

$$\sum_{i=1}^{m} n_i - \sum_{i=1}^{m} \frac{1+\alpha^2(n_i-1)}{(1+\alpha(n_i-1))^2} e_i' T_1(\rho_1) \otimes T_2(\rho_2) e_i = 0. \tag{6.13}$$

A method such as bisection can be applied to obtain a solution $g(\rho_1, \rho_2)$ (for α) to Equation (6.13).

The stage one estimator $\hat{\alpha}_{QONE}$ therefore converges in probability to $g(\rho_1, \rho_2)$ when the true Kronecker product correlation structure is misspecified as exchangeable. When the true structures are assumed to be exchangeable, the stage two estimate $\hat{\alpha}_{QLS}$ is obtained as the solution $f(\alpha)$ to the stage two estimating Equation (3.41) that is evaluated at $\alpha = \hat{\alpha}_{QONE}$ and exchangeable structures $R_i(\alpha)$. Because

$\widehat{\alpha}_{QONE} \xrightarrow{p} g(\rho_1, \rho_2)$ when the true Kronecker product structure is misspecified as exchangeable, it then follows that the limiting value of the stage two estimate $\widehat{\alpha}_{QLS}$ converges in probability to $f(g(\rho_1, \rho_2))$. The limiting value of $\widehat{\alpha}_{QLS}$ can therefore be obtained as the solution for α to (3.41) with $g(\rho_1, \rho_2)$ substituted for $\widehat{\alpha}_{QONE}$.

The stage two estimating equation for the exchangeable working structure has a closed form solution that was provided in Equation (5.8). We can then therefore apply the following algorithm to obtain the limiting value of $\widehat{\alpha}_{QLS}$ when the true Kronecker product structure has been misspecified as exchangeable:

(1) For assumed true values of ρ_1, ρ_2, use the bisection method to obtain a solution $g(\rho_1, \rho_2)$ (for α) to Equation (6.13).

(2) Next, obtain the limiting value of $\widehat{\alpha}_{QLS}$ by evaluating Equation (5.8) with $g(\rho_1, \rho_2)$ substituted for $\widehat{\alpha}_{QONE}$.

Chapter 7

Correlated Binary Data

In this chapter we focus on binary data, which has been the topic of some recent debate in the statistical literature. For example, Ziegler and Vens (2010); Breitung et al. (2010); Molenberghs (2010), and Shults (2011) all focus on the special issues related to binary data; these authors were specifically interested in the selection of correlation structures for dichotomous outcomes (which we will turn to in the next chapter), but they also touched on some of the special features of binary data that complicate the use of GEE and QLS for their analysis.

A fundamental problem of GEE and QLS for analysis of binary data stems from their semi-parametric nature, which is often described as a beneficial feature of these methods. In particular, GEE and QLS specify separate models for the marginal mean and working correlation structure; this simplifies their implementation because there is no need to fully specify the underlying multivariate distribution of the data. However, relative to continuous outcomes, there are additional constraints that must be satisfied by the regression and correlation parameters for discrete data in order for a valid multivariate distribution to exist. Checks for satisfaction of these constraints are not built into the standard software packages for GEE, so that any violation is likely to go undetected. Violation of constraints for dichotomous outcomes is therefore unlike a negative estimate of variance for continuous outcomes that is also invalid but is easy to detect.

In this chapter we review some of the recent discussion that relates to GEE and QLS analysis of binary data. We also include a comparison between QLS and the first-order Markov maximum-likelihood (MARK1ML) approach that was developed by Guerra et al. (2012). Maximum-likelihood is considered by many to be the "gold standard"; our demonstration that QLS (and GEE) hold up well relative to MARK1ML in many situations is therefore helpful to alleviate concerns raised in the statistical literature (e.g., by Chaganty and Mav (2007)) that QLS and GEE may not be appropriate for the analysis of binary data. Those who are interested in learning more about the development and application of MARK1ML are encouraged to consult Guerra and Shults (2014); this text also contains additional technical details regarding the issues that we discuss in this chapter.

In addition to MARK1ML we also describe some comparisons with alternating logistic regression (ALR). ALR is widely used for analysis of correlated binary data because it models associations via odds ratios, which are subject to constraints that are less restrictive than the constraints for the correlations (Chaganty and Joe, 2006, Section 5). ALR can therefore be applied as a means of circumventing the issues related to constraints for the correlations for binary data; however, this might not be sufficient to justify selecting ALR over competing approaches if ALR does not have superior performance with respect to other aspects of the analysis.

7.0.1   Notation for Correlated Binary Data

In this chapter we use some special notation for discrete random variables. Let $pr(y_{ij} = a)$ represent the probability that y_{ij} takes value a. In addition, let $pr(y_{ij} = 1)$ = p_{ij} and $q_{ij} = 1 - p_{ij}$.

When discussing the correlation between two random variables, we continue to use the same notation that we described in Section 2.3.4. We also continue to assume that we have correctly specified the correlation structure, so that the working structure $R_i(\alpha)$ and the true structure $T_i(\rho)$ are identical. However, we also consider situations in which the true structure is misspecified, so that the QLS and GEE estimators of the correlation parameter do not necessarily converge in probability to the true correlation parameter ρ. It is also helpful to let $R_i[j,k]$ represent the $(j,k)^{th}$ element of $R_i(\alpha)$, which is the assumed correlation between y_{ij} and y_{ik}. In other words we model $Corr(y_{ij}, y_{ik})$ by $R_i[j,k]$, where $R_i[j,k]$ is the $(j,k)^{th}$ element of $R_i(\alpha)$.

7.1   Additional Constraints for Binary Data

For ease of understanding we describe the additional constraints for binary data in the context of a toenail study that was described in Section 1.5.3 of Chapter 1. For this demonstration we restrict our attention to the 224 patients with complete data (117 in group A and 104 in group B).

```
. use toenail, clear
. sort id time
. by id: keep if _N==7
```

We next fit a QLS model using **xtqls** in Stata (Shults et al., 2007) to compare the two treatment groups over time with a logistic model for severe toenail infection that includes time, treatment group, and a time by treatment interaction term as covariates. We specify an exchangeable working correlation structure and display the estimated correlation matrix after fitting the logistic model with QLS.

```
. quietly xtqls y time treatment interaction, i(id) t(time) f(bin 1) c(exc)
  vce(robust)

(output suppressed)

. xtcorr

Estimated within-id correlation matrix R:

          c1       c2       c3       c4       c5       c6       c7
r1    1.0000
```

```
r2  0.4269  1.0000
r3  0.4269  0.4269  1.0000
r4  0.4269  0.4269  0.4269  1.0000
r5  0.4269  0.4269  0.4269  0.4269  1.0000
r6  0.4269  0.4269  0.4269  0.4269  0.4269  1.0000
r7  0.4269  0.4269  0.4269  0.4269  0.4269  0.4269  1.0000
```

We obtained our results without receiving any warning that there is a potential problem with respect to the estimated parameters. If we fit GEE, the results would be similar, with no indication that the estimates are potentially invalid. (See Exercise 7.1.) However, there is a hidden problem that we will uncover. We start by obtaining the fitted values:

```
-> . predict phat
(option mu assumed; Pr(y != 0))
```

An important feature of dichotomous random variables is that their expected values represent the probability of a positive response (see Exercise 7.2), which is indicated in the output following implementation of the **predict** command above. Because the fitted values are identical within cells defined by treatment group and measurement occasion for our assumed model, we can easily display the predicted probabilities for the two treatment groups at each time point by using the **table** command in Stata to display the sample mean of the estimated marginal means by treatment group and time:

```
-> . table treatment time, c(mean phat)
```

	time						
treatment	0	1	2	3	6	9	12
0	.3434795	.29639835	.25328097	.21452353	.12478081	.06926981	.03739865
1	.34380826	.28665079	.23558414	.19117687	.09635532	.0458955	.02123962

According to the above table the probability of a severe toenail infection is approximately 0.34 for both groups at time 0 and then declines to less than 0.04 for both groups at time 12.

7.1.1 Negative Estimated Bivariate Probabilities for the Toenail Data

With some additional simple calculations we now demonstrate the problem associated with the analysis results in Section 7.1. We do this by first noting that even though we usually use QLS and GEE to estimate the probability that a patient has the outcome of interest at a particular measurement occasion, the estimated parameter values that we obtain from QLS or GEE *induce* all the bivariate probabilities as a byproduct of the analysis. For example, the estimates of the correlation and regression parameters completely determine the probability that a patient in treatment group 0 has no infection at baseline, but does have an infection at the last measurement occasion. We run into trouble when one of the induced probabilities is negative, as we now demonstrate for the toenail study.

Given the estimates $\widehat{\alpha}$ and $\widehat{\beta}$ we can obtain the estimated probabilities $pr(y_{ij} =$

$a, y_{ik} = b)$, where $(a,b) \in \{(0,1),(1,0),(0,0),(1,1)\}$, using the following expression for the bivariate Bernoulli distribution that was provided in Prentice (1988). (The following formula is easy to verify, as seen in Exercises 7.3 and 7.4.)

$$\text{pr}(y_{ij} = a, y_{ik} = b) =$$

$$p_{ij}^a q_{ij}^{1-a} p_{ik}^b q_{ik}^{1-b} \left\{ 1 + Corr(y_{ij}, y_{ik}) \frac{(a - p_{ij})(b - p_{ik})}{(p_{ij} q_{ij} p_{ik} q_{ik})^{1/2}} \right\}. \qquad (7.1)$$

For our analysis of the toenail data we assumed an exchangeable working structure, so that we modeled $Corr(y_{ij}, y_{ik})$ as $R_i[j,k] = \alpha$ for all i, j, and k in Equation (7.1). According to the output displayed earlier, $\widehat{\text{pr}(y_{i1} = 1)} = 0.3434795$, $\widehat{\text{pr}(y_{i7} = 1)} = 0.0373986$, and $\widehat{\alpha} = 0.4269$ for all patients in group 0. We next substitute these values into Equation (7.1) and simplify the resultant expression slightly, in order to obtain an estimate of $\text{pr}(y_{11} = 0, y_{17} = 1)$:

```
. scalar alpha = 0.4269
. scalar p11 = 0.3434795
. scalar p17 = 0.0373986
. scalar probab = (1-p11)*p17
                - alpha*sqrt((1-p11)*p17*p11*(1-p17))
. scalar list probab
     probab =  -.0139108
```

The estimated bivariate probability $\text{pr}(y_{11} = 0, y_{17} = 1)$ is therefore equal to -0.0139108. The QLS (and GEE) estimates of the marginal means and pairwise correlations would therefore force us to say that the probability that a patient in group zero has no infection at the first measurement occasion but does have an infection at the final visit is *negative*; therefore, the joint distribution of y_{11} and y_{17} is invalid and there is no valid underlying multivariate distribution that is compatible with the assumed model and its estimated parameters.

7.1.2   *Prentice Constraints to Ensure Valid Induced Bivariate Distributions*

It is easy to check that the four probabilities defined by Equation (7.1) all sum to one; however, as demonstrated in the previous section, in some situations the analysis results yield a negative bivariate probability, in which case the bivariate Bernoulli distribution that is induced by a particular QLS (or GEE) analysis is invalid.

Solving $\text{pr}(y_{ij} = a, y_{ik} = b) \geq 0$ for all $(a,b) \in \{(0,1),(1,0),(0,0),(1,1)\}$ and $\text{pr}(y_{ij} = a, y_{ik} = b)$ as defined in Equation (7.1) yields the following constraints on the pairwise correlations and marginal means that must be satisfied in order to achieve a valid bivariate Bernoulli distribution: (See Exercises 7.3 and 7.4.)

$$L_i(j,k) \leq Corr(Y_{ij}, Y_{ik}) \leq U_i(j,k), \qquad (7.2)$$

where $\quad L_i(j,k) \quad = \quad \max\left\{-(w_{ij}w_{ik})^{1/2}, -(w_{ij}w_{ik})^{-1/2}\right\}, \quad U_i(j,k) \quad =$

$\min\left\{ (w_{ij}/w_{ik})^{1/2}, (w_{ij}/w_{ik})^{-1/2} \right\}$, and $w_{ij} = p_{ij}(1-p_{ij})^{-1}$, for $i = 1,\ldots,m$ and $j = 1,\ldots,n_i - 1$. We refer to Equation (7.2) as the Prentice constraints, which Prentice (1988) described while discussing GEE models.

When we conduct a QLS or GEE analysis, we obtain estimates of the marginal means p_{ij} and pairwise correlations $Corr(y_{ij}, y_{ik})$. If the constraints (7.2) are not satisfied for a particular $\widehat{p}_{ij}$, $\widehat{p}_{ik}$, and $\widehat{Corr}(y_{ij}, y_{ik})$, then the induced bivariate Bernoulli distribution between a particular y_{ij} and y_{ik} will be invalid. Because $\widehat{Corr}(y_{ij}, y_{ik})$ is a function of the estimator α, we can obtain the interval $(\widehat{L}, \widehat{U})$ on which $\widehat{\alpha}$ satisfies the constraints for all $\widehat{p}_{ij}$, $\widehat{p}_{ik}$ and $\widehat{Corr}(y_{ij}, y_{ik})$. We refer to $\widehat{L}$ and $\widehat{U}$ as the *estimated* Prentice constraints for $\widehat{\alpha}$ (or to L and U as the Prentice constraints for α when the parameter values are known).

Continuing with the example considered in Section 7.1.1, we next obtain the Prentice constraints using the user-written Stata command **checkPrenticeequi** that takes the following arguments: id, time, phat (estimated probabilities), and the estimated value of α, all of which are needed to check satisfaction of the inequalities in (7.2).

```
.  checkPrenticeequi id phat time 0.4269
```

```
The estimated value of alpha for the exchangeable working
structure is .4269
```

```
The estimated Prentice constraints are (-.03230891 , .20351313).
The Prentice constraints are violated for the exchangeable
structure, which means that there is no underlying multivariate
distribution that corresponds with the fitted model.
```

In Section 7.1.1 we calculated a negative estimated bivariate Bernoulli probability, which indicated that the Prentice constraints were violated for the toenail data. We now see that the violation is severe because the estimated $\widehat{\alpha}$ of 0.4269 is considerably larger than the estimated upper constraint of 0.2035.

In general, a difficulty in the typical application of GEE and QLS is that checks for violation of constraints are not built into the currently available software for implementation of QLS and GEE. An exception is the SAS macro that was developed by Hammill and Preisser for GEE that is available at the following website:

http://www.bios.unc.edu/~preisser/personal/geediag/diag103.sas.

In addition, Kim and Shults (2010) developed software for QLS that provides a warning if the Prentice constraints are violated for binary data. The latest version of the **xtqls** command, **xtqls2** (Shults) also automatically checks for a violation of the Prentice constraints, as we demonstrate by again fitting the model that we implemented in Section 7.1.1.

```
.  xtqls2 y time treatment interaction, i(id) t(time) f(bin 1) c(exc) vce(robust)
```

```
Iteration 1: tolerance = .02340574
Iteration 2: tolerance = .00498732
Iteration 3: tolerance = .00089506
```

```
Iteration 4: tolerance = .00004521
Iteration 5: tolerance = 5.062e-07
```

```
GEE population-averaged model                    Number of obs      =      1568
Group and time vars:              id __00001C    Number of groups   =       224
Link:                                   logit    Obs per group: min =         7
Family:                              binomial                   avg =       7.0
Correlation:              fixed (specified)                     max =         7
                                                 Wald chi2(3)       =     50.15
Scale parameter:                          1      Prob > chi2        =    0.0000
```

```
                                   (Std. Err. adjusted for clustering on id)
-----------------------------------------------------------------------------
             |               Semirobust
         y   |     Coef.    Std. Err.      z    P>|z|     [95% Conf. Interval]
-------------+---------------------------------------------------------------
        time |  -.2166815    .0428493   -5.06   0.000    -.3006645   -.1326985
   treatment |   .0014576    .3078827    0.00   0.996    -.6019814    .6048966
 interaction |  -.0486559    .0690008   -0.71   0.481    -.183895     .0865831
       _cons |  -.6478265    .2087196   -3.10   0.002    -1.056909   -.2387435
-----------------------------------------------------------------------------
```

```
The estimated value of alpha  is .42689867
```

```
The estimated Prentice constraints are (-.03230891 , .20351313).
The Prentice constraints are violated,
which means that there is no underlying multivariate distribution that
corresponds with the assumed model.
```

The recently updated **xtqls2** command (Shults) therefore automatically obtained the estimated Prentice constraints and provided a warning that they are violated for the exchangeable working structure.

7.1.3 Simplification of the Prentice Constraints for Decaying Product Correlation Structures

When we conduct a QLS or GEE analysis we obtain estimates of the marginal means p_{ij} and pairwise correlations $Corr(y_{ij}, y_{ik})$. In the previous sections we demonstrated that for some fitted values $\hat{p}_{ij}$ and estimated correlations $\widehat{Corr(y_{ij}, y_{ik})}$, the induced bivariate Bernoulli distribution between a particular y_{ij} and y_{ik} will be invalid unless the estimates satisfy the Prentice constraints (7.2) for all $\hat{p}_{ij}$, $\hat{p}_{ik}$, and $\widehat{Corr(y_{ij}, y_{ik})}$. However, the following theorem indicates that for some working correlation structures it is not necessary to check that all the pairwise inequalities are satisfied in (7.2) in order to confirm satisfaction of the Prentice constraints for all $\hat{p}_{ij}$, $\hat{p}_{ik}$, and $\widehat{Corr(y_{ij}, y_{ik})}$; for a particular class of correlation structures it will be sufficient to check that the consecutive pairs $\hat{p}_{ij}$, $\widehat{p_{ij+1}}$, and $\widehat{Corr(y_{ij}, y_{ij+1})}$ satisfy the constraints for $j = 1, \cdots, n_i - 1$ and $i = 1, \cdots, m$.

Theorem 7.1 (Prentice Constraints for Decaying Product Structures) *Suppose we consider a decaying product working correlation structure, for which $Corr(y_{ij}, y_{ik})$ is modeled by $\prod_{w=j}^{k-1} R_i[w, w+1]$; then satisfaction of the Prentice constraints (7.2) for* *the adjacent correlations $Corr(y_{iw}, y_{iw+1}) = R_i[w, w+1]$ for all i and $w = 1, \cdots, k-1$ is sufficient to ensure satisfaction of the Prentice constraints for $\prod_{w=j}^{k-1} R_i[w, w+1]$. Therefore for decaying product correlation structures, which include the AR(1) structure (for $R_i[w, w+1] = \alpha$) and the Markov structure (for $R_i[w, w+1] = \alpha^{t_{iw+1} - t_{iw}}$), satisfaction of the Prentice constraints for the adjacent correlations $R_i[w, w+1]$ for $w = j, \cdots, k-1$ is sufficient to guarantee satisfaction of the constraints for $R_i[j, k]$.*

Proof 1 (Proof of Theorem 7.1) *The following is a very slight generalization of a proof that was provided in Kim et al. (2008a) for the AR(1) correlation structure. Without loss of generality assume that $j < k$. We first consider the upper bounds. Let*

$$
\begin{aligned}
m &= \min\{U_i(j, j+1), U_i(j+1, j+2), \ldots, U_i(k-1, k)\} \\
&= \min\{v_{ij}, (v_{ij})^{-1}, v_{ij+1}, (v_{ij+1})^{-1}, \ldots, v_{ik-1}, (v_{ik-1})^{-1}\},
\end{aligned}
$$

where $v_{ij} = (w_{ij}/w_{ij+1})^{1/2}$. Then satisfaction of the Prentice constraints (7.2) for the adjacent correlations $R_i[w, w+1]$ implies that $\prod_{w=j}^{k-1} R_i[w, w+1] \leq m^{k-j}$, where

$$
\begin{aligned}
m^{k-j} &\leq v_{ij}v_{ij+1}\ldots v_{ik-1} = (w_{ij}/w_{ik})^{1/2}, \text{ and} \\
m^{k-j} &\leq (v_{ij})^{-1}(v_{ij+1})^{-1}\ldots(v_{ik-1})^{-1} = (w_{ij}/w_{ik})^{-1/2}.
\end{aligned}
$$

Therefore $\prod_{w=j}^{k-1} R_i[w, w+1] \leq \min\{(w_{ij}/w_{ik})^{1/2}, (w_{ij}/w_{ik})^{-1/2}\} = U_i(j, k)$. Next, we consider the lower bounds. The lower bounds will be satisfied when $k-j$ is even because in this case $\prod_{w=j}^{k-1} R_i[w, w+1] > 0$ and the lower bound is always negative. We therefore only need to consider the case that $k-j$ is odd. Let

$$
\begin{aligned}
s &= \max\{L_i(j, j+1), L_i(j+1, j+2), \ldots, L_i(k-1, k)\} \\
&= \max\{-z_{ij}, -(z_{ij})^{-1}, -z_{ij+1}, -(z_{ij+1})^{-1}, \ldots, -z_{ik-1}, -(z_{ik-1})^{-1}\},
\end{aligned}
$$

where $z_{ij} = (w_{ij}w_{ij+1})^{1/2}$. Then Equation (7.1) and the fact that $k-j$ is odd implies that $\prod_{w=j}^{k-1} R_i[w, w+1] \geq s^{k-j}$, where

$$
\begin{aligned}
s^{k-j} &\geq -z_{ij}(z_{ij+1})^{-1}z_{ij+2}\ldots(z_{ik-2})^{-1}z_{ik-1} = -(w_{ij}w_{ik})^{1/2}, \text{ and} \\
s^{k-j} &\geq -(z_{ij})^{-1}z_{ij+1}(z_{ij+2})^{-1}\ldots z_{ik-2}(z_{ik-1})^{-1} = -(w_{ij}w_{ik})^{-1/2}.
\end{aligned}
$$

Therefore $\prod_{w=j}^{k-1} R_i[w, w+1] \geq \max\{-(w_{ij}w_{ik})^{1/2}, -(w_{ij}w_{ik})^{-1/2}\} = L_i(j, k)$. Because our results do not depend on the particular choice of i, j, and k, the result of the theorem follows.

Using the fact that the Prentice constraints only need to be checked for the adjacent correlations for the AR(1) and Markov structures, it is easy to simplify the constraints for these structures. Table 7.1 that was provided in Shults (2011) expresses the constraints (7.2) for a logistic model as a function of the estimated regression and correlation parameters, and is particularly useful to calculate the Prentice constraints on the basis of estimates that are provided in a journal article.

Table 7.1 *(From Shults (2011).) Prentice Constraints for* α, *for a Logistic Model and Several Working Correlation Structures. Vector* x_{ij} *is the* $p \times 1$ *vector of covariates measured on subject i at measurement occasion j, while* β *is the* $p \times 1$ *vector of regression coefficients. Note that there can be additional constraints that must be satisfied, in order to obtain a positive definite correlation matrix. For the exchangeable structure,* α *must take value in* $(-1/n_m, 1)$, *where* n_m *is the maximum value of the number of measurements* n_i *on subject i. For the tri-diagonal structure,* α *must take value in* $(-1/c_m, 1/c_m)$, *where* $c_m = 2 \sin\left(\frac{\pi[n_m-1]}{2[n_m+1]}\right)$.

	Exchangeable	Tri-diagonal & AR(1)
$\mathbf{L_W}$	$\max_{i,j,k}\left\{-exp(-\lvert(x_{ij}+x_{ik})'\beta\rvert/2)\right\}$	$\max_{i,j}\left\{-exp(-\lvert(x_{ij}+x_{ij+1})'\beta\rvert/2)\right\}$
$\mathbf{U_W}$	$\min_{i,j,k}\left\{exp(-\lvert(x_{ij}-x_{ik})'\beta\rvert/2)\right\}$	$\min_{i,j}\left\{exp(-\lvert(x_{ij}-x_{ij+1})'\beta\rvert/2)\right\}$

We next return to the toenail data and continue with the estimation of the Prentice constraints. In Section 7.1.2 we demonstrated that there was a severe violation of constraints for the exchangeable working structure. We next estimate the Prentice constraints for the Markov correlation structure with the user-written command **checkPrenticeMarkov** (Shults):

```
*Next, check that constraints for the Markov structure.
. xtqls y time treatment interaction, i(id) t(time) f(bin 1) c(Markov) vce(robust)

(output suppressed)

. xtcorr

Estimated within-id correlation matrix R:

          c1      c2      c3      c4      c5      c6      c7
r1    1.0000
r2    0.8278  1.0000
r3    0.6852  0.8278  1.0000
r4    0.5672  0.6852  0.8278  1.0000
r5    0.3217  0.3887  0.4695  0.5672  1.0000
r6    0.1825  0.2205  0.2663  0.3217  0.5672  1.0000
r7    0.1035  0.1251  0.1511  0.1825  0.3217  0.5672  1.0000

. predict phat2
(option mu assumed; Pr(y != 0))

. checkPrenticeMarkov id phat2 time 0.8278
The estimated value of alpha  is .8278

The estimated Prentice constraints are (-.29223423 , .88201247).
```

Therefore, there is no violation of the Prentice constraints for the Markov structure, while there was a severe violation for the exchangeable structure. (Or, as an alternative to fitting **xtqls** followed by **checkPrenticeMarkov** the command **xtqls2** automatically checks and reports a violation of constraints.) Shults et al. (2009) viewed a severe violation of constraints as an indication that the correlation structure has been misspecified. Violation of constraints could also indicate a problem with the model for the marginal mean, as demonstrated in Kim and Shults (2010). We shall return to these issues of goodness of fit and selection of a working correlation structure in a subsequent chapter.

7.1.4   Conditions to Ensure the Existence of an Underlying Multivariate Distribution

It is important to note a caveat regarding the calculation of the Prentice constraints; if they are not satisfied, there can be no valid multivariate distribution with the estimated parameter values for the assumed model. If they are satisfied for adjacent correlations, this is sufficient to guarantee the existence of a multivariate distribution for a *decaying product* correlation structure, namely the first-order Markov (MARK1) model that was implemented for maximum-likelihood analysis in Guerra et al. (2012) and that is thoroughly explored in Guerra and Shults (2014).

However, if the working structure is not of decaying product form, then satisfaction of the constraints may not be sufficient to guarantee the existence of an underlying multivariate distribution (parent distribution) with the assumed parameter values. For example, Chaganty and Joe (2006) presented a simple example for a tri-diagonal correlation structure, for which the Prentice constraints were satisfied but for which there was no valid parent distribution. Qaqish (2003) proved that satisfaction of the Prentice constraints (which he termed natural restrictions) for adjacent correlations is sufficient to guarantee the existence of a valid conditional linear distribution for the AR(1) structure; he also provided a general approach for obtaining the restrictions on the correlations that are sufficient to guarantee the existence of a valid conditional linear parent distribution.

Molenberghs and Kenward (2010) use a hybrid conditional model to construct a parent distribution for given marginal means and correlations; their approach rewrites the correlation-based bivariate probabilities in odds-ratio form. Molenberghs and Kenward (2010) also provide an example for which there is no valid Bahadur (1961) parent distribution, but there is a valid hybrid conditional model. As noted by Molenberghs and Kenward (2010), the relevance of their construction "lies in the fact that the parent provides a natural description of the framework into which the semi-parametrically specified parameters fit. The implication is that such semi-parametric methods as GEE1, GEE2, ALR, etc. can always be applied because there is always a valid parent, and hence a probabilistic basis. The sole condition is that the parametrically specified portion of the model be valid, but this is no different to any other statistical modeling exercise."

7.2   When Violation Is Likely to Occur

Because there can be no valid parent distribution if the Prentice constraints are violated, it is useful to examine the conditions under which a violation is likely to occur. We continue with the scenario considered earlier in this chapter, in which we compared the probability of a positive response between two treatment groups over time via the following model:

$$\text{pr}(y_{ij} = 1) = \frac{\exp{(\delta_{ij})}}{1 + \exp{(\delta_{ij})}}, \tag{7.3}$$

where $\delta_{ij} = x'_{ij}\boldsymbol{\beta}$ and $\delta_{ij} = \beta_0 + \beta_1 I(groupB) + \beta_2 visit + \beta_3 I(groupB) \times visit$; $visit = 0, 1, \cdots, n$; and $I(groupB) = 1$ for subjects in group B (and 0 for subjects in group A). The number of subjects in each group is $m/2$. The simulations were conducted using the approach described in Guerra and Shults (2013) for overdispersed discrete data; this method reduces to the approach proposed by Qaqish (2003) and the same model considered by Zeger et al. (1985), for longitudinal Bernoulli data with an AR(1) structure.

7.2.1   When the Model Is Correctly Specified

If the models for the mean and correlation structures are correctly specified, then the regression and correlation parameters will be estimated consistently and the Prentice constraints will be satisfied asymptotically. For finite sample sizes, simulations suggest that violation of the Prentice constraints will be unlikely, unless the sample size is extremely small or α is close to the lower or upper boundary values in (7.2). Figure 7.1 displays the proportion of 1,000 simulation runs that GEE analysis (with virtually identical results for QLS) that resulted in violation of the Prentice constraints for model (8.7) when the correlation structure was correctly specified as AR(1) and $\boldsymbol{\beta} = (-1.3862944, 0, 1.2862944, -1.1167961)'$; this demonstrates that even for a very small sample size of 20 subjects per group, the proportion of simulations that resulted in a violation of the Prentice constraints was low unless α was close to a boundary value. In addition, as the sample size increased, there was less likely to be a violation of the Prentice constrains for all values of α; the *severity* of any violation also declined with increasing group size. For example, Figure 7.2 indicates that for $\alpha = 0.49$ (which was very close to the upper Prentice constraint $U = 0.50$ in Figure 7.1), the estimated correlation $\hat{\alpha}$ tended to be very close to the estimated Prentice constraint $\hat{U}$. In other words, for a moderate sample size violation of the constraints was unlikely unless α was very close to a boundary value, in which case a *mild violation* of constraints was more likely to occur.

Additional simulations for a true exchangeable structure yielded similar results as for the correctly specified AR(1) structure. When the true and working structures were exchangeable and the group sizes are 30 or more, any violation of constraints was unlikely unless α was very close to the boundary value. In addition, as for the AR(1) structure, any observed violation of the estimated Prentice constraints tended to be mild. In general, a severe violation of constraints was highly unlikely for a correctly specified correlation structure.

7.2.2   When the Working Structure Is Incorrectly Specified

However, if the model is misspecified, then the bounds may be violated asymptotically and the violation may be severe. For example, suppose we consider the scenario in which the model for the mean is correct but the true AR(1) correlation structure is

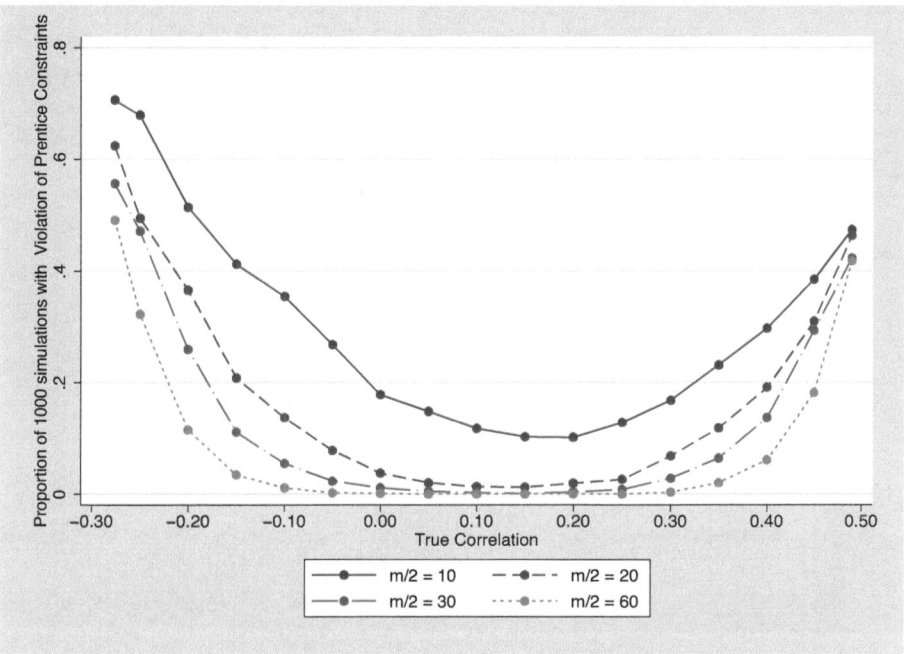

Figure 7.1 *Proportion of 1,000 simulations that resulted in a violation of the Prentice constraints versus true α for model (8.7) when the working structure is correctly specified as AR(1). Probabilities of a positive response are (0.2, 0.5, 0.8) and (0.2, 0.2466, 0.30) at visits 1, 2, and 3, for groups A and B, respectively; $\boldsymbol{\beta} = (-1.3862944, 0, 1.2862944, -1.1167961)'$. The Prentice constraints for α (7.2) are $(L, U) = (-0.2861, 0.50)$.*

misspecified as exchangeable. In this situation the upper and lower bounds for α will be estimated consistently, but $\hat{\alpha}$ will not be consistent, with a limiting value that can be obtained using the algorithm described in Section 6.8 of Chapter 6. The limiting values will not necessarily agree for GEE and QLS when the true AR(1) structure is misspecified as exchangeable; however, if the number of measurements is the same for all subjects, then the limiting value for both GEE and QLS can be expressed as the average of the pairwise correlations (from (B.7) of Shults et al. (2006b)):

$$limit = \frac{2\sum_{j=1}^{n-1}\sum_{k=j+1}^{n}\rho^{k-j}}{(n-1)n}, \tag{7.4}$$

where ρ is the true value of the correlation parameter and n is the number of measurements per subject.

In the previous section we considered an example for which the probabilities of a positive response are (0.2, 0.5, 0.8) and (0.2, 0.2466, 0.30) at visits 1, 2, and 3, for groups A and B; the Prentice constraints (7.2) for this example are $(-0.29, 0.25)$ for the exchangeable correlation structure (versus $(-.29, 0.50)$ for the AR(1) structure). When the AR(1) structure is misspecified as exchangeable, the QLS (or GEE) esti-

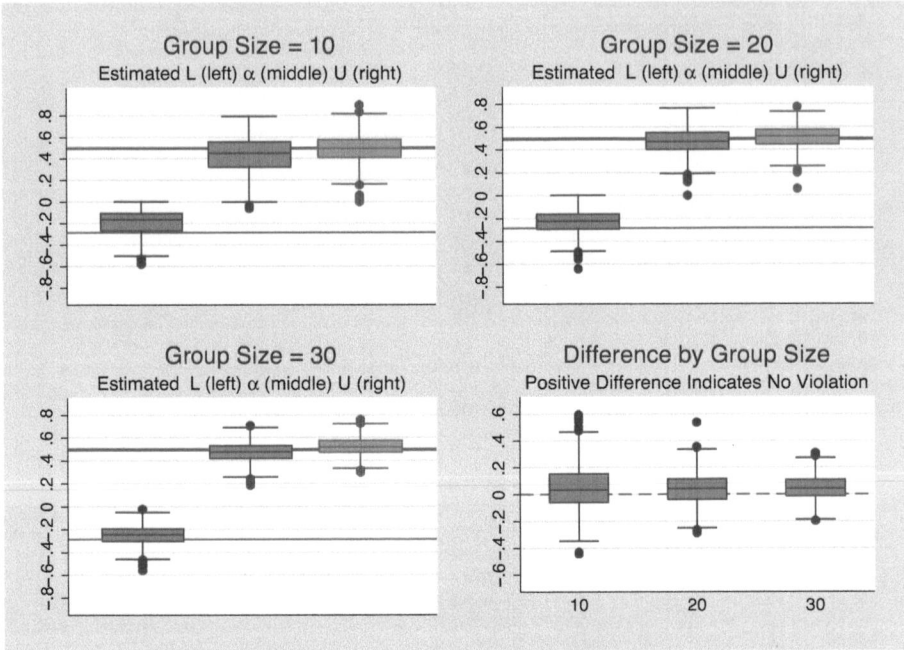

Figure 7.2 *Box plots of $\widehat{L}$ (left), $\widehat{\alpha}$ (center), $\widehat{U}$ (right) in 1,000 simulation runs for the same model as in Figure 7.1 when $\alpha = 0.49$ and the true AR(1) structure is correctly specified. Here α is very close to the upper Prentice constraint U. The true values of L, α, and U are -0.2861, 0.49, and 0.50, respectively; Horizontal lines are displayed at each of these values. Group sizes are 10 (upper-left graph), 20 (upper-right graph), and 30 (lower-left graph). The lower-right graph displays boxplots of the difference, $\widehat{U} - \widehat{\alpha}$, between the estimated Prentice constraint and correlation for each group size. (There is no violation of the upper Prentice constraint if the difference is positive.) As the sample size increases, the estimates of the constraints (L,U) and of α become closer to their true values; In addition, the difference becomes smaller, so that any violation of the Prentice constraints tends to become less severe with increasing sample size.*

mator of the upper value of the Prentice constraint will tend in probability to the true upper limit of 0.25 for the exchangeable structure; however, the corresponding estimator of the correlation parameter will not tend in probability to the true correlation parameter ρ. If we use Equation (7.4) to obtain the limiting value for $n = 3$, we obtain

$$limit = \frac{2\rho + \rho^2}{3}. \qquad (7.5)$$

The limiting value in (7.5) will exceed the limiting value of the upper Prentice constraint when

$$\frac{2\rho + \rho^2}{3} > 0.25,$$

which will occur when $\rho > 0.323$. The asymptotic violation of constraints can also

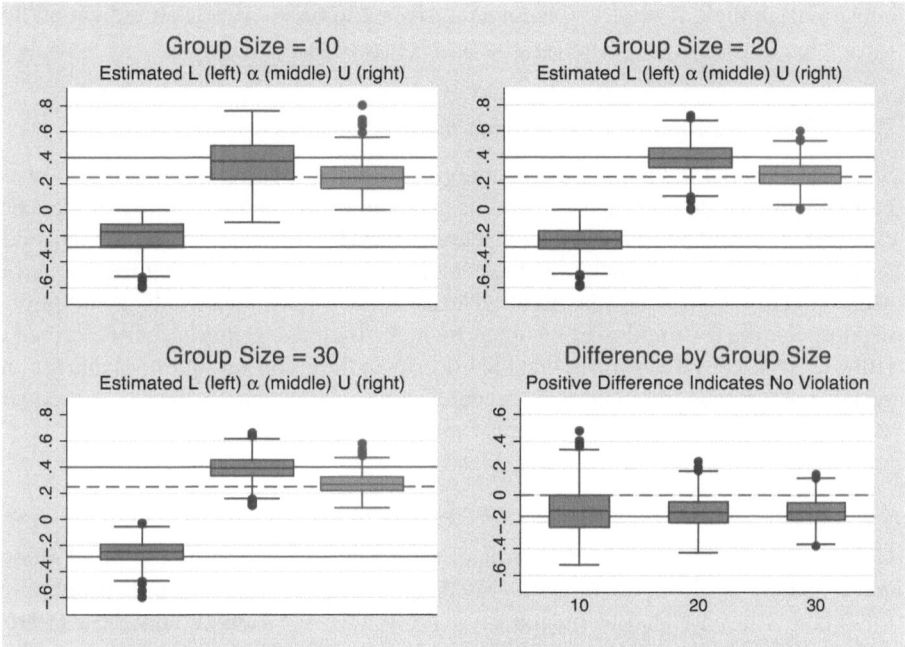

Figure 7.3 *Boxplots of $\widehat{L}$ (left), $\widehat{\alpha}$ (center), $\widehat{U}$ (right) in 1,000 simulation runs for the same model as in Figure 7.1 when $\alpha = 0.49$ and the true AR(1) structure is misppecified as exchangeable. The true values of L, U, and the limiting value of $\widehat{\alpha}$ are -0.2861, 0.25, and 0.4067, respectively; horizontal lines are displayed at each of these values. Group sizes are 10 (upper-left graph), 20 (upper-right graph), and 30 (lower-left graph). The lower-right graph displays boxplots of the difference, $\widehat{U} - \widehat{\alpha}$, between the estimated prentice constraint and correlation for each group size. (There is no violation of the upper Prentice constraint if the difference is positive.) As the sample size increases the estimates of the constraints (L,U) and of α become closer to their true values; in addition, the difference approaches its limiting value, $U - \alpha = 0.25 - 0.4067 = -0.1567$. Therefore, this demonstrates that the severity of violation of bounds does not necessarily become less severe as the sample size increases, when the working structure is exchangeable, but the true structure is AR(1).*

be quite severe; for example, if $\rho = 0.49$, then the QLS (or GEE) estimator of ρ converges in probability to 0.4067, which is quite a bit larger than the upper constraint of 0.25. Figure 7.3 indicates that for finite samples when $\rho = 0.49$, the estimated correlation $\hat{\alpha}$ did substantially exceed the estimated Prentice constraint $\hat{U}$; the median violation became more severe (more negative) with increasing sample size, as the estimates became closer to their limiting values.

In another example, row 3 of Table 1 in Shults et al. (2009) displays a scenario for which the limiting value of the upper boundary value for the Prentice constraints is 0.1682 but the limiting value of $\hat{\alpha}$ is 0.5089, when the true AR(1) structure (with $\rho = 0.7752$) has been misspecified as exchangeable; this was a severe violation of constraints, which, if observed in an analysis, could be viewed as an indication that the choice of working correlation structure is incorrect.

7.2.3   When the Model for the Mean Is Incorrect

Suppose next that the model for the marginal mean is incorrect; this can also result in a violation of the Prentice constraints, asymptotically and for finite samples. For example, Kim and Shults (2010) considered a model for analysis of the toenail study (see Section 1.5.3) for which both the exchangeable and AR(1) structures resulted in a violation of the Prentice constraints. The authors assessed their model assumptions, in particular the assumption of the linearity in the logit using approaches described in Hilbe (2009). They determined that the linearity assumption was questionable for one variable. They then modified their model accordingly, which resulted in subsequent satisfaction of the Prentice constraints for these data.

7.2.4   When the Assumption of Missing Completely at Random Is Violated

GEE and QLS yield unbiased parameter estimates when any missing observations are missing completely at random (MCAR), which means that the distribution of missingness is unrelated to outcomes or covariates; see Heitjan and Basu (1996), Rubin (1976), Little and Rubin (2002) and Hedeker and Gibbons (2006) for further discussion of missing data. Troxel et al. (1998) provided a thorough description of issues related to missing data in the analysis of quality of life data in clinical trials. If the distribution of missingness is related to observed outcomes or covariates, then the data are missing at random (MAR) and the estimating equations for the regression parameter will not necessarily be unbiased. As a result, the GEE and QLS estimators will not necessarily be consistent if the distribution of missingness is MAR.

Preisser et al. (2002) assessed a weighted estimating equation approach that weights observations or person-visits in the GEE estimating equation for β with weights that are inversely proportional to their probability of being observed. They performed extensive simulations to show that the weighted GEE approach reduces small sample bias relative to GEE when the drop-out mechanism is correctly specified; however, misidentification of the drop-out mechanism could result in worse performance for the weighted GEE method relative to GEE. Troxel et al. (2010) presented a pseudo-likelihood approach for analysis of longitudinal binary data subject to non-ignorable, non-monotone missingness. Guerra et al. (2012) proposed the MARK1ML approach for analysis of longitudinal binary data that is appropriate for data with a MAR missingness mechanism. MARK1ML will be discussed in Section 7.4.

7.3   Implications of Violation of Constraints for Binary Data

Violation (and in particular severe violation) of the Prentice constraints could suggest a problem with model assumptions regarding the model for the mean and working correlation structure; we will explore this more thoroughly in a subsequent chapter that considers model fit and choice of working correlation structure. In general however, we suggest that a severe violation of Prentice constraints should be viewed as an indication that the choice of working correlation structure is incorrect, if we have confidence in our model for the marginal mean.

Another potential area of importance regarding violation of constraints relates to sample size calculations for future studies. Sample size formulae for longitudinal studies require specification of values for the marginal means and correlation parameter, but will not provide a warning if values are proposed that violate the Prentice constraints. It is therefore possible to base the proposed sample sizes for a study on assumed parameter values that do not correspond to any true multivariate distribution and which are therefore impossible values if the assumed model is correct. As Rochon (1998) cautioned, "the practitioner must be aware of these restrictions (the Prentice constraints), particularly at the design stage."

7.4 Comparison among GEE, QLS, and MARK1ML

As mentioned earlier in this chapter, there has been some suggestion in the recent statistical literature that semi-parametric approaches such as QLS and GEE may not be appropriate for the analysis of binary data. A comparison among QLS, GEE, and an ML approach for analysis of binary data is therefore very informative. ML-based approaches that yield estimates that correspond with a valid parent distribution are appropriate under parametric modeling assumptions, so that if GEE and QLS perform well in comparison with such an ML-based approach, this should ease concerns regarding their appropriateness for the binary data of interest.

The ML approach we consider is the MARK1ML method that was developed by Guerra et al. (2012). MARK1ML is based on the likelihood that is obtained under the first-order Markov (MARK1) assumption that the distribution of each y_{ij} only depends on the immediately antecedent y_{ij-1}. Maximum-likelihood is used for estimation, so that the MARK1 assumption combined with ML for estimation yields the "MARK1ML" approach. However, Guerra et al. (2012) also proved that the estimating equations for *both the regression and correlation parameters* will be unbiased, as long as the correlations between the adjacent y_{ij} and y_{ij+1} are correctly specified (which is equivalent to correctly specifying the off-diagonal elements of the true correlation structure). As a result, MARK1ML is also robust to misidentification of the true correlation structure, although not to the same extent as GEE and QLS, which involve estimating equations for the *regression parameter* that are unbiased even if the adjacent correlations are misspecified.

Guerra et al. (2012) made extensive comparisons among MARK1ML, GEE, ALR, and QLS; however, the results for QLS were not shown because the findings were very similar for QLS and GEE. Table 7.2 displays the MSE for MARK1ML, GEE, and QLS for Model 8.7 with $\beta = (1, 0.1, -0.2, -0.3)$. Table 7.2 shows that the MSE is very slightly larger for QLS and GEE than for MARK1ML when the working correlation structure is correctly specified as AR(1). For group sizes $m = 120$ (results not shown), the MSEs were smaller for all methods, and the MSE was again only slightly smaller for MARK1ML in comparison with QLS and GEE. Table 7.2

Table 7.2 *MSE for β and α for MARK1ML, GEE, and QLS in* 5000 *Simulation Runs for Model (8.7) when $\beta = (1, 0.1, -0.2, -0.3)$ and $\alpha \in \{0.25, 0.50, 0.75\}$. The true and working correlation structures are both AR(1).*

m	α	Method	β_0	β_1	β_2	β_3	α
60	0.25	MARK1ML	0.1138	0.2291	0.0110	0.0243	0.0035
60	0.25	GEE	0.1150	0.2324	0.0111	0.0247	0.0036
60	0.25	QLS	0.1150	0.2322	0.0111	0.0247	0.0036
60	0.50	MARK1ML	0.1428	0.2882	0.0119	0.0259	0.0027
60	0.50	GEE	0.1475	0.3013	0.0121	0.0271	0.0029
60	0.50	QLS	0.1474	0.3011	0.0121	0.0271	0.0029
60	0.75	MARK1ML	0.1633	0.2897	0.0096	0.0165	0.0014
60	0.75	GEE	0.1719	0.3425	0.0100	0.0228	0.0015
60	0.75	QLS	0.1720	0.3429	0.0100	0.0229	0.0015

then displays the MSE when QLS and GEE are implemented with an AR(1) working structure, but the true correlation structure is exchangeable. Table 7.2 shows that the methods are very similar when the true exchangeable structure is misspecified as AR(1), which is as expected because all approaches have unbiased estimating equations for β in this scenario. When the group size was increased to $m = 120$ (results not shown), the MSE decreased for all methods, as expected.

The results of the simulations presented here suggest that although GEE and QLS suffer a slight loss in small-sample efficiency relative to the MARK1ML approach, the difference among approaches is not serious enough to suggest that GEE and QLS are not appropriate for analysis of binary data. Although not shown here, the percentage bias was also estimated for all scenarios, with similar findings as for the MSE. The coverage probabilities based on a sandwich covariance matrix were also obtained for the regression parameters; close to nominal 95% levels were obtained for all scenarios and approaches.

However, there is one situation for which there will be a meaningful difference between approaches. Guerra et al. (2012) demonstrated via extensive simulations that MARK1ML offers important improvements in performance relative to GEE and ALR when the data are missing at random (MAR); this was as anticipated because while ML-based approaches are appropriate if the data are MAR, GEE and QLS require the more restrictive assumption that data are missing completely at random (MCAR).

7.4.1 Comparisons with ALR

In addition to the results presented in the previous section, Guerra et al. (2012) made extensive comparisons via simulation with ALR, using the ORTH package in R (By et al., 2011). We describe their findings here. These additional comparisons were important because ALR models association via odds ratios that are not subject to the same restrictive constraints as are correlation parameters. However, although the odds ratios are not as severely restricted as are correlations, the simulations indicated

Table 7.3 *MSE for β and α for MARK1ML, GEE, and QLS in 5000 Simulation Runs for Model (8.7) When $\beta = (1, 0.1, -0.2, -0.3)$ and $\alpha = 0.25$. The true exchangeable correlation structure is misspecified as AR(1).*

m	α	Method	β_0	β_1	β_2	β_3	α
60	0.25	MARK1ML	0.1090	0.2073	0.0067	0.0145	0.0052
60	0.25	GEE	0.1112	0.2127	0.0068	0.0150	0.0053
60	0.25	QLS	0.1111	0.2126	0.0068	0.0150	0.0052
120	0.25	MARK1ML	0.0532	0.1008	0.0033	0.0069	0.0025
120	0.25	GEE	0.0541	0.1039	0.0033	0.0072	0.0025
120	0.25	QLS	0.0541	0.1039	0.0033	0.0072	0.0025

some serious issues with ALR in terms of failure to converge and increased bias and MSE for larger correlations; the authors conjectured that ALR might have difficulties when the odds ratios are very large, which is the case for higher values of the correlation. Therefore, although ALR models associations with odds ratios that are subject to less restrictive constraints than the constraints for correlations (Chaganty and Joe, 2006), its performance could suffer when the odds ratios are very large; however, this suspicion admittedly requires further investigation.

Simulations were conducted when the true correlation structure was correctly specified, and also when only the off-diagonal elements of the true structure were correctly specified, for GEE, ALR, MARK1ML, and ALR and exchangeable, AR(1), Markov, or first-order antedependent correlation structures. However, for ALR, the true induced correlation structure was not correctly specified in most scenarios because ALR requires an assumption regarding the pattern of values in the odds ratios that does not necessarily correspond to the same pattern in the correlations. For example, if we assume that all the odds ratios are equal, so that the pattern of association in the odds ratios is exchangeable, then this induces a pattern of association in the correlations that is *not* exchangeable, except for the trivial no-covariates case. Therefore, if we simulate data with exchangeable correlations and fit GEE with an exchangeable working correlation structure and ALR with a working assumption that the odds ratios are exchangeable, then the correlation structure will be correctly specified for GEE, but incorrectly specified for ALR. Therefore, the comparisons with ALR were conducted under conditions that were perhaps more favorable to the competing approaches (with the exception of some simulations with a true unstructured matrix), although one might argue that the assumed working correlation structures represented reasonable patterns of association that one might expect to encounter in longitudinal studies. As a result, the simulations that showed superior performance of MARK1ML and GEE to ALR were meaningful because they were made under realistic scenarios for longitudinal trials.

7.5 Prentice-Corrected QLS and GEE

The simulations that were described in the previous sections ignored any violation of the Prentice constraints for QLS and GEE. However, QLS (and GEE) can be adjusted so that the final estimates satisfy the additional Prentice restrictions for the correlations. First, fit unadjusted GEE or QLS with working structure $R_i(\alpha)$ to obtain $\widehat{\beta}$ (that determines $\widehat{p}_{ik}$), $\widehat{\alpha}$, and the interval $(\widehat{L}, \widehat{U})$ on which $\widehat{\alpha}$ satisfies the constraints (7.2) for all $\widehat{p}_{ij}$, $\widehat{p}_{ik}$ and $\widehat{Corr(y_{ij}, y_{ik})}$. (This interval can be obtained for several common correlation structures and a logistic model using Table 7.1.) If $\widehat{\alpha}$ violates the Prentice constraints because $\widehat{\alpha} > \widehat{U}$, then do the following. Set a tolerance level, for example, $tolerance = 0.00001$. Let $count = 1$, $j = 1$, and $upper = \widehat{\alpha}$.

   Step 1: Let $\widehat{\alpha} = upper - \frac{1}{10}^{count} \times j$.

   Step 2: Update the estimate $\widehat{\beta}$ of β by solving the GEE estimating equation evaluated at $R_i(\widehat{\alpha})$. Update the estimated Prentice constraints $(\widehat{L}, \widehat{U})$ and check if $\widehat{\alpha}$ satisfies the updated estimates of the Prentice constraints (7.2). If not, that is, if $\widehat{\alpha} > \widehat{U}$, then let $j = j + 1$ and repeat Steps 1 and 2. If $\widehat{L} \leq \widehat{\alpha} \leq \widehat{U}$ but $tolerance < \frac{1}{10}^{count}$, then let $upper = \widehat{\alpha} + \frac{1}{10}^{count}$; then $j = 1$; $count = count + 1$; and repeat Steps 1 and 2. If $\widehat{L} \leq \widehat{\alpha} \leq \widehat{U}$ and $tolerance \geq \frac{1}{10}^{count}$, then the process is done and the adjusted estimate $\widehat{\alpha}_{ADJ}$ is given by $\widehat{\alpha}$. The adjusted estimate $\widehat{\beta}$ is the current solution to the GEE estimating equation for β.

   If $\widehat{\alpha}$ violates the estimated Prentice constraints because $\widehat{\alpha} < \widehat{L}$, then the above algorithm can be applied, for $lower = \widehat{\alpha}$, $\widehat{\alpha} = lower + \frac{1}{10}^{count} \times j$ and $lower = \widehat{\alpha} - \frac{1}{10}^{count}$ substituted for their corresponding values for $upper$ in Steps 1 and 2, respectively.

   Simulations (not shown) conducted in Shults et al. (2006b) show that when the model is correctly specified, the adjusted QLS (or GEE) approach offers almost no improvements in terms of MSE and bias in estimation of the regression and correlation parameters in comparison with an unadjusted approach, when the correlation is in the mid-range of the Prentice constraints. Why is this? When α is far from the Prentice constraints, violation of the constraints is extremely unlikely, so that adjustment is usually not necessary for smaller values of the correlation; adjustment is rarely implemented and thus has little impact. Implementing the adjustment can offer a very small improvement when the true value of α is close to a boundary value of the interval (L, U) on which α satisfies the constraints in (7.2); in this situation, a mild violation of constraints is likely and adjusting the final estimate of α to force satisfaction of the Prentice constraints offered some very minor improvements in terms of MSE and bias, when compared with ignoring any potential violation.

   However, Prentice-corrected QLS can offer meaningful improvements if estimation of the bivariate probabilities (7.1) is desired and the correlation is large and close to a Prentice constraint; in this situation, violation of the constraints is more likely to occur, in which case some of the estimated bivariate probabilities will be negative. Shults et al. (2006b) compared adjusted QLS with an approach proposed by Chaganty and Joe (2004). Chaganty and Joe (2004) suggested that in GEE analysis of binary data, an exchangeable structure should be implemented for "cluster-type samples," while a first-order autoregressive AR(1) structure should be applied for longitudinal data. The correlation parameter α could then be estimated with a value $\widehat{\alpha}_r$ that is close to zero (for weak dependence), in the range 0.2 to 0.3 (for moderate dependence), or in the range 0.4 to 0.7 (for strong dependence), where "small," "medium," and "large" are defined on the basis of descriptive analyses that include assessment of odds ratios. Alternatively, Chaganty and Joe (2004) suggested that α could be estimated with the midpoint $\hat{\alpha}_m$ of the Prentice constraints for an exchangeable working correlation structure; we refer to this as the mid-range approach. The regression parameter would then be estimated by solving the GEE estimating equation evaluated at $\hat{\alpha}_r$ or $\hat{\alpha}_m$.

Shults et al. (2006b) showed that application of the mid-range approach versus adjusted QLS or GEE can result in severe asymptotic bias in estimation of the correlation parameter and of the bivariate probabilities. Table 1 in Shults et al. (2006b) presents simulation results that show that application of the mid-point approach versus GEE and QLS can also result in biased and inefficient estimation of α and of the bivariate probabilities for small samples when α is large. Results were very similar for QLS and GEE. Shults et al. (2009) also implemented the approach of Chaganty and Joe (2004), and suggested that the method could be difficult to apply in practice because it is somewhat arbitrary with respect to the selection of a final estimate of the correlation parameter.

7.6 Summary

For dichotomous outcomes, the correlations are subject to additional constraints. Satisfaction of the Prentice constraints for α is necessary to guarantee the existence of a valid multivariate parent distribution that provides a probabilistic basis for semi-parametric approaches such as QLS and GEE. For decaying product correlation structures (e.g., AR(1), Markov), satisfaction of the Prentice constraints is sufficient to guarantee the existence of a valid parent distribution.

Unlike the failure to obtain a positive-definite estimated correlation matrix, failure to satisfy the Prentice constraints will not typically result in a failure to converge for QLS and GEE. As a result, violation of the Prentice constraints often goes undetected in a GEE or QLS analysis. However, although it is possible to implement algorithms (e.g., as in the previous section) that force satisfaction of the Prentice constraints, we suggest fitting unadjusted QLS or GEE first and then checking for a violation. Prentice-adjusted QLS or GEE might be applied when there is a very minor violation (for example, $(\hat{L}, \hat{U}) = (-0.202, 0.315)$ and $\hat{\alpha} = 0.320$); a very slight change to the final estimate of α is unlikely to have a major impact on the results, but it will allow for valid estimation of the bivariate probabilities (no negative bivariate probability estimates).

No or mild violation of the Prentice constraints does not prove that the model fits well; however, *a severe violation is extremely unlikely for a moderate sample size and a model that has been correctly specified*. Therefore, we suggest that a severe violation of the Prentice constraints should prompt a careful assessment of the model assumptions. For example, if confidence in the model for the marginal mean is high, then a severe violation could be viewed as a strong indication that the choice of working correlation structure is incorrect. Further discussion of issues related to goodness of fit and use of a severe violation of constraints to rule out particular correlation structures will be provided in the next chapter.

7.7   Exercises

Exercise 7.1 *Replicate the analysis provided at the start of Section 7.1 for GEE in Stata, to demonstrate that the results are similar for GEE and QLS for analysis of the toenail data that is restricted to subjects with complete data.*

Exercise 7.2 *Suppose that random variable y_{ij} is Bernoulli with $pr(y_{ij} = 1) = p_{ij}$ and $pr(y_{ij} = 0) = 1 - p_{ij}$. Show that the expected value of y_{ij} is equal to p_{ij}. Also show that the variance of y_{ij} is equal to $p_{ij}(1 - p_{ij})$. (Hint: A very useful feature of Bernoulli random variables is that $(y_{ij})^k = y_{ij}$ for $y_{ij} \in \{0, 1\}$.)*

Exercise 7.3 *Consider Bernoulli random variables y_{ij} and y_{ik} with expected values p_{ij} and p_{ik} and correlation $Corr(Y_{ij}, Y_{ik})$.*

    a. Use the definition of correlation to obtain $Pr(Y_{ij} = 1, Y_{ik} = 1)$ as a function of the expected values and correlation.

    b. Obtain the remaining bivariate probabilities $Pr(Y_{ij} = 0, Y_{ij+1} = 1)$, $Pr(Y_{ij} = 1, Y_{ik} = 0)$, and $Pr(Y_{ij} = 0, Y_{ik} = 0)$. (Hint: The bivariate probabilities can be calculated easily using $Pr(Y_{ij} = a, Y_{ik} = 0) + Pr(Y_{ij} = a, Y_{ik} = 1) = Pr(Y_{ij} = a)$ and $Pr(Y_{ij} = 0, Y_{ik} = b) + Pr(Y_{ij} = 1, Y_{ik} = b) = Pr(Y_{ik} = b)$.)

    c. Confirm that the bivariate probabilities will always sum to one.

Exercise 7.4 *Verify that Equation 7.1 yields the same probabilities that you obtained in problem 7.3 for the bivariate Bernoulli distribution.*

Exercise 7.5 *Verify directly that the constraints provided in (7.2) must be satisfied for the correlation between any two Bernoulli random variables. (Hint: The constraints can be obtained directly by obtaining the conditions under which the pairwise probabilities that you obtained in Exercise 7.3 are non-negative.)*

Exercise 7.6 *Find the Prentice constraints (7.2) for the model considered in Figure 7.1 that represented a comparison of two groups for which the probabilities of a positive response are (0.2, 0.5, 0.8) and (0.2, 0.2466, 0.30) at visits 1, 2, and 3, for groups A and B, respectively.*

Chapter 8

Assessing Goodness of Fit and Choice of Correlation Structure for QLS and GEE

QLS models are evaluated as to fit based on the same basis as GEE models. In general, assessing fit is challenging for semi-parametric methods such as QLS and GEE, as these approaches are not based on maximizing an objective function. In contrast, maximum likelihood-based approaches estimate the regression parameter by maximizing the log-likelihood, whose estimated value can then be used to compare the fit of different models. For example, the Akaike information criterion (AIC) is widely used for assessing the relative fit of models based on maximum likelihood (Akaike, 1973):

$$AIC = 2p - 2ln(L_{\hat{\beta}_{ML}}),$$ (8.1)

where p is the dimension of the regression parameter β, $\hat{\beta}_{ML}$ is the maximum likelihood estimate of β, and $ln(L_{\hat{\beta}})$ is the natural-log of the likelihood evaluated at $\hat{\beta}_{ML}$. The AIC criterion can be used to compare several candidate models, assuming the proposed model includes the true model, by designating the model with the smallest AIC as the model with the best relative fit. The AIC values will tend to be smaller if the number of parameters in the model is smaller, and if the estimated log-likelihood is larger. The AIC is therefore favorably influenced by models that are more parsimonious, and that have greater estimated log-likelihoods.

However, the use of AIC can lead to severe over-fitting, so that the dimension of the regression parameter in the fitted model is too large. Hurvich and Tsai (1989) therefore built on previous work by Sugiura (1978) to develop a correction to the AIC criterion. Although the AIC was designed to be an approximately unbiased estimator of the expected Kullback–Leibler information of a fitted model, Hurvich and Tsai (1989) noted that "As m, the dimension of the candidate model, increases in comparison to n, the sample size, AIC becomes a strongly negatively biased estimate of the information." They therefore suggested the following bias corrected version of the AIC (Hurvich and Tsai, 1989, Equation (4) on p. 300):

$$AIC_C = AIC + \frac{2(M+1)(M+2)}{n-M-2}$$ (8.2)

where M is the dimension of the proposed model and n is the sample size. They considered regression models with independent errors, for which M represented the dimension of the regression parameter, and autoregressive models of order M. We note that Equation (8.2) is identical to Equation (4) of Hurvich and Tsai (1989), with one small change; we replaced the lowercase m with an uppercase M because in this book we use m to denote the number of subjects.

While Sugiura (1978) originally proposed a bias correction for AIC for linear regression models, Hurvich and Tsai (1989) used simulations to assess the small sample performance of AIC_C for nonlinear regression and time-series models. The second term in Equation (8.2) tends to zero as the sample size increases, so that the potential impact of the bias correction is greater for small samples. Hurvich and Tsai (1989) suggest that the correction will be especially useful for small sample sizes, or when the number of parameters in the candidate model is a moderate to large fraction of the sample size.

Pan (2001) extended the AIC statistic for the semi-parametric GEE approach using quasi-likelihood constructed under the working independence model and model-based and robust sandwich covariance estimates of the estimated regression coefficients. The fit test he designed, referred to as the quasi-likelihood information criterion (QIC), can be used for covariate model selection or to determine the most appropriate correlation structure for a given GEE or QLS model. The QIC criterion is defined as

$$QIC = -2L_{I,\hat{\boldsymbol{\beta}}_{\hat{R}_i}} + 2trace(\hat{\Sigma}^{-1}_{M,I,\hat{\boldsymbol{\beta}}_{\hat{R}_i}} \hat{\Sigma}_{S,\hat{R}_i,\hat{\boldsymbol{\beta}}_{\hat{R}_i}}), \qquad (8.3)$$

where $L_{I,\hat{\boldsymbol{\beta}}_{\hat{R}_i}}$ is the quasi-likelihood calculated using the identity working correlation structure, $\hat{\boldsymbol{\beta}}_{\hat{R}_i}$ is the estimate of $\boldsymbol{\beta}$ obtained under the assumed working correlation structure R_i, $\hat{\Sigma}_{S,\hat{R}_i,\hat{\boldsymbol{\beta}}_{\hat{R}_i}}$ is the sandwich-based covariance matrix defined in Equation (2.24) that is evaluated at the working structure R_i and at $\hat{\boldsymbol{\beta}}_{\hat{R}_i}$, and $\hat{\Sigma}_{M,I,\hat{\boldsymbol{\beta}}_{\hat{R}_i}}$ is the model-based covariance matrix defined in Equation (2.25) but evaluated now for an **identity** working correlation structure and at $\hat{\boldsymbol{\beta}}_{\hat{R}_i}$. Note that in this chapter we slightly modify our notation, for example by including additional subscripts in our names of estimators; we do this to distinguish more easily between the different goodness of fit criteria that have been defined in the statistical literature. Hardin and Hilbe (2003) defined the QIC as

$$QIC_I = -2L_{I,\hat{\boldsymbol{\beta}}_{\hat{R}_i}} + 2trace(\hat{\Sigma}^{-1}_{M,I,\hat{\boldsymbol{\beta}}_I} \hat{\Sigma}_{S,\hat{R}_i,\hat{\boldsymbol{\beta}}_{\hat{R}_i}}), \qquad (8.4)$$

where $\hat{\Sigma}_{M,I,\hat{\boldsymbol{\beta}}_I}$ is the model-based covariance matrix evaluated for the identity working correlation structure and at $\hat{\boldsymbol{\beta}}_I$, where $\hat{\boldsymbol{\beta}}_I$ is obtained under assumption of the **identity** working correlation structure.

Of two GEE, or QLS, models with calculated QIC statistics, the model with a lower QIC value is the preferred, or better fitted, model. Starting with version 9.2 of SAS, the QIC criterion is available in **PROC GENMOD**. SAS also provides $QIC_u = QIC + p$, where p is the dimension of $\boldsymbol{\beta}$. The QIC_u therefore includes a penalty term for models that are less parsimonious; like the *AIC* criterion, it is favorably influenced by simpler models. The QIC_u is recommended for choosing between models for the mean (which need not be nested) but not for selecting an appropriate correlation structure.

Zheng (2000) details several goodness-of-fit tests for population-averaged panel models in general, which include GEE, GEE2, ALR, and QLS. However, none has found widespread support in software applications. Grady and Helms (1995) propose graphical techniques that can be used to visually assess patterns in the estimated correlations.

Horton et al. (1999) devised a goodness-of-fit test for binomial GEE models based on the well-known Hosmer–Lemeshow test for binary logistic regression. The idea of the test is to model the data, predict fitted values and rank, them into ten deciles of risk. The values within each decile are compared to their associated ranked observed values. A Chi-squared goodness-of-fit statistic is used to compare the observed versus predicted counts within and over deciles. The problem with this test is that tied values can account for many, if not most, of both the predicted and observed counts. We suggest employing the test with eight, ten, and twelve panels of risk, making certain that the resultant Chi-squared statistic is similar in value for each test.

The enhanced Hosmer–Lemeshow test is not appropriate for testing the adequacy of QLS correlation structures, but it is useful for evaluating the functional form of the model and as a specification test for the predictors used in the model. Keeping these caveats in mind, the test is a useful check on whether the data are structured properly or if there are a number of observations in the data that unduly influence other parts.

The *QIC* criterion has remained a popular method for selection of a working correlation structure for GEE models since it was initiated in 2001. Cui (2007) provided software for implementation of the *QIC* criterion in Stata. Hin and Wang (2009) ran a series of simulations demonstrating that the *QIC* statistic is biased toward the unstructured correlation. The true structures were special cases of the unstructured matrix in the simulation scenarios considered by Hin and Wang (2009); the tendency to choose an unstructured matrix therefore did not represent a tendency to choose an incorrect structure in their simulations. However, Westgate (2013) showed that estimating multiple correlation parameters in the unstructured matrix can increase the variances of the regression parameters in small samples, so that it might be preferable to fit a simpler structure when the simpler structure is correctly specified.

Barnett et al. (2010) ran additional simulations in which they compared a deviance information criterion (*DIC*, using Bayesian models), with the *AIC* (using mixed models) and *QIC* and QIC_I criteria (using GEE) with respect to their ability to choose between an AR(1), exchangeable, identity, and unstructured correlation matrix. They used the **MIXED** procedure in SAS to fit mixed models with restricted maximum likelihood, the **GENMOD** procedure in SAS to implement GEE, and **WinBUGS** to fit Bayesian models (Spiegelhalter et al., 2002). Barnett et al. (2010) observed little difference in performance between the *QIC* and QIC_I criteria, and therefore referred solely to one *QIC* criterion in the later part of their manuscript.

In our own simulations we have observed that the QIC and QIC_I criteria are virtually indistinguishable, and we therefore focus on the QIC criterion in this chapter.

Barnett et al. (2010) concluded that the AIC statistic is in fact a better criterion for selection of a working correlation structure than is the QIC. However, Barnett et al. (2010) ran their simulations for data that were distributed according to a multivariate normal distribution. As shown in Park (1993), GEE is equivalent to a maximum likelihood approach for balanced multivariate normal data with an unstructured correlation matrix. It is therefore perhaps not that surprising that the maximum likelihood-based AIC outperforms the QIC for normal data.

Barnett et al. (2010) also noted that the original simulations that were conducted by Pan (2001) did not include the unstructured correlation among the candidate working structures; had the unstructured correlation been included, the simulations in Barnett et al. (2010) indicated that the success rates for QIC would have been lower. In fact, Barnett et al. (2010) suggested that "the QIC is untrustworthy, and should not be used for selecting among competing covariance structures." Of all the criteria they compared, Barnett et al. (2010) suggested the use of the DIC criterion.

Hardin and Hilbe (2012) suggested adapting Hurvich's adjusted AIC criterion (8.2) for the QIC criterion; we take a similar approach here. We replace the dimension of the candidate model M in (8.2), with $p + r$, to include both the dimension p of $\boldsymbol{\beta}$ and the number of correlation parameters r in the dimension of the approximating model. We also replace the sample size n, with the number of subjects m, to obtain the following adjusted QIC criterion:

$$QIC_A = QIC + \frac{2(p+r+1)(p+r+2)}{m-p-r-2}. \qquad (8.5)$$

Hardin and Hilbe (2012) reported simulation results that demonstrated that penalizing the QIC criterion for a greater dimension of the correlation parameter reduced the likelihood of selecting an unstructured matrix over a simpler structure, but also resulted in a tendency to incorrectly select the identity structure for smaller sample sizes.

Hin and Wang (2009) studied both terms in the QIC criterion in Equation (8.3). The first term in (8.3) does not involve the working correlation structure, although the estimates of $\boldsymbol{\beta}$ are obtained at the assumed structure. They conducted simulations that showed that the first term in Equation (8.3) is more sensitive to changes in the model for the marginal mean than to incorrect specification of the working correlation structure. To select a working correlation structure for GEE, Hin and Wang (2009) suggest using half of the second term of the QIC, which they refer to as the correlation information criterion, or CIC:

$$CIC = 2trace\left(\hat{\Sigma}_{M,I,\hat{\boldsymbol{\beta}}_I}^{-1} \hat{\Sigma}_{S,\hat{R}_i,\hat{\boldsymbol{\beta}}_{\hat{R}_i}}\right). \qquad (8.6)$$

Hilbe (2009) recommended using both the CIC and QIC, with the preferred model to be one in which both statistics have lower values than any competing model, when such a result is obtained.

Additional criteria for selection of a working correlation structure are the

Rotnitzky–Jewell (*RJ*) criteria, which compare the model-based and sandwich-based estimates of the covariance matrix of $\hat{\boldsymbol{\beta}}$ under the assumed working correlation structure $\hat{R}_i$. If the assumed working correlation structure is correct, then the model-based and sandwich-based estimates of the covariance matrix should be similar, that is,

$$\hat{\Sigma}_{M,\hat{R}_i,\hat{\beta}_{\hat{R}_i}} \approx \hat{\Sigma}_{S,\hat{R}_i,\hat{\beta}_{\hat{R}_i}},$$

so that

$$Q = \hat{\Sigma}_{M,\hat{R},\hat{\beta}_{\hat{R}_i}}^{-1} \hat{\Sigma}_{S,\hat{R}_i,\hat{\beta}_{\hat{R}_i}} \approx I$$

and

$$Q^2 \approx I.$$

The Rotnitzky–Jewell (*RJ*) criteria (Rotnitzky and Jewell, 1990) are defined as

$$
\begin{aligned}
RJ1 &= trace(Q)/p, \\
RJ2 &= trace(Q^2)/p, \text{ and} \\
DBAR &= \sum_j (e_j - 1)^2 = RJ2 - 2RJ1 + 1,
\end{aligned}
$$

where the e_j are the eigenvalues of Q, and p (the dimension of $\boldsymbol{\beta}$) is also the dimension of Q. When the assumed working correlation structure is correct, we would therefore expect that $RJ1 \approx 1$, $RJ2 \approx 1$, and $DBAR \approx 0$. We therefore selected the working correlation structure for which $|RJ1 - 1|$, $|RJ2 - 1|$, or $|DBAR|$ is smallest, for the $RJ1$, $RJ2$, and $DBAR$ criteria, respectively. The RJ criteria are attractive because they are intuitively reasonable for selection of a working correlation structure.

Shults et al. (2009) compared the performance of the *RJ* criteria with several other criteria for selection of a correlation structure for GEE analysis of correlated binary data, including the Shults–Chaganty (SC) criterion (Shults and Chaganty, 1998). Shults et al. (2009) showed that the SC criterion has relatively poor performance for correct selection of the working correlation structure.

White (2011) conducted simulations to compare the *RJ*, *CIC*, and *QIC* for selection of a working correlation structure for QLS for multivariate normal data, when the true and working correlation structures were exchangeable, tri-diagonal, AR(1), and Markov; he observed that none of the methods did a good job distinguishing between the AR(1) and Markov correlation structures. White (2011) also proposed and evaluated a criterion based on the Frobenius matrix norm. As demonstrated by Hin and Wang (2009), White (2011) also observed that the *CIC* outperformed the *QIC* criterion with respect to selecting the true correlation structure.

Shults et al. (2009) did not evaluate the *CIC*, *QIC*, or *QIC*$_A$ criteria in their assessment of methods for selection of a working correlation structure for GEE analysis of binary data. In this chapter we extend their comparisons to include these additional criteria; however, we do not evaluate the SC criterion that was considered by these authors, due to its poor performance in their simulations.

8.1   Simulation Scenarios

We used the method described in Online Appendix B (Section 2.1) of Guerra et al. (2012) and in Guerra and Shults (2013) to simulate longitudinal binary data; the method stems from the model in Zeger et al. (1985) and is related to the approach of Qaqish (2003) that is demonstrated in Preisser et al. (2002).

We compared the *CIC*, *QIC*, *QIC_A* and the three *RJ* criteria with respect to their ability to correctly identify the true correlation structure for QLS. In addition to assessing the individual performance of each criterion by the percentage of times the correct structure was selected, we also evaluated the number of times that each working structure was chosen by three or more criteria. In practice, being selected by several criteria might support the selection of a particular structure. In addition, we evaluated the number of times a structure was selected by no criteria, because this might be viewed as convincing evidence that a particular structure is not correct. Simulation results for QLS and GEE were virtually identical, so that we only show the results for QLS here.

We considered the following model

$$\mathrm{pr}(y_{ij} = 1) = \frac{\exp(\delta_{ij})}{1 + \exp(\delta_{ij})}, \tag{8.7}$$

where $\delta_{ij} = x'_{ij}\boldsymbol{\beta}$ and $\delta_{ij} = \beta_0 + \beta_1 I(groupB) + \beta_2 time + \beta_3 I(groupB) \times time$; $time \in \{t_{i1}, \ldots, t_{i6}\}$, and $I(groupB) = 1$ for subjects in group B (and 0 for subjects in group A). The number of subjects in each group is $m/2$, where $m = 138$.

The true regression parameter $\boldsymbol{\beta} = (\beta_0, \beta_1, \beta_2, \beta_3) = (2.122, -.617, -.3, 0.101)'$. The true correlation structures (see Theorem 7.1) were AR(1) with $\alpha = 0.5$, Markov with $\alpha = 0.5$, and a decaying product (DP) correlation structure (see Theorem 7.1) with adjacent correlations $R[1,2] = .20$; $R[2,3] = .30$; $R[3,4] = .40$; $R[4,5] = .45$; and $R[5,6] = 0.50$. For the true AR(1) structures, we assume that the subjects had a common set of measurement times, and that the consecutive measurements were equally spaced on each subject. For the true Markov structure, we assume that the timing of measurements varied between subjects; 46 subjects (all in group A) had timings = $(1,2,3,4,5,6)$; 46 (23 in each group) had timings = $(1,4,8,12,16,20)$; and 46 (all in group B) had timings = $(1,2,3,6,9,12)$.

The true correlation structures can be expressed as follows.

The true AR(1) structure is

$$\begin{pmatrix}
1 & 0.5 & 0.25 & 0.125 & 0.0625 & 0.03125 \\
0.5 & 1 & 0.5 & 0.25 & 0.125 & 0.0625 \\
0.25 & 0.50 & 1 & 0.5 & 0.25 & 0.125 \\
0.125 & 0.25 & 0.5 & 1 & 0.5 & 0.25 \\
0.0625 & 0.125 & 0.25 & 0.5 & 1 & 0.5 \\
0.03125 & 0.0625 & 0.125 & 0.25 & 0.5 & 1
\end{pmatrix}.$$

The unstructured correlation structure includes the AR(1) as a special case; therefore, if the true structure is AR(1) it would not be incorrect to fit an unstructured matrix. However, implementation of an unstructured matrix would involve estimation of fifteen correlation parameters, versus one for the AR(1) structure; as a result, we could suffer some small sample loss in efficiency if we fit the more general unstructured matrix, even though it is not incorrect when the true structure is AR(1).

Next, the true Markov structures depend on the spacing of measurements. For the 23 subjects with timings timings $= (1,2,3,4,5,6)$, the true structure is the AR(1) structure above. For the 23 subjects with timings $= (1,4,8,12,16,20)$, the true Markov structure is:

$$
\begin{pmatrix}
1 & 0.5 & 0.25 & 0.03125 & 0.00390625 & 0.00048828 \\
0.5 & 1 & 0.5 & 0.0625 & 0.0078125 & 0.00097656 \\
0.25 & 0.5 & 1 & 0.125 & 0.015625 & 0.00195313 \\
0.03125 & 0.0625 & 0.125 & 1 & 0.125 & 0.015625 \\
0.00390625 & 0.0078125 & 0.015625 & 0.125 & 1 & 0.125 \\
0.00048828 & 0.00097656 & 0.00195313 & 0.015625 & 0.125 & 1
\end{pmatrix}.
$$

For the 92 subjects with timings $= (1,2,3,6,9,12)$, the true Markov structure is

$$
\begin{pmatrix}
1 & 0.125 & 0.0078125 & 0.00048828 & 0.00003052 & 10.907e-06 \\
0.125 & 1 & 0.0625 & 0.00390625 & 0.00024414 & 0.00001526 \\
0.0078125 & 0.0625 & 1 & 0.0625 & 0.00390625 & 0.00024414 \\
0.00048828 & 0.00390625 & 0.0625 & 1 & 0.0625 & 0.00390625 \\
0.00003052 & 0.00024414 & 0.00390625 & 0.0625 & 1 & 0.0625 \\
10.907e-06 & 0.00001526 & 0.00024414 & 0.00390625 & 0.0625 & 1
\end{pmatrix}.
$$

Assuming an AR(1), exchangeable, identity, or unstructured matrix would not be correct for the true assumed Markov structures, because these working structures all incorrectly assume that the correlation structure is the same for all subjects. However, the AR(1) structure is correct for a subset of twenty-three subjects. (In general, the correlation structure will not be the same for all subjects when the Markov structure is the true structure and the temporal spacing of measurements is not the same for all subjects.)

Finally, the decaying product correlation (DP) structure has the following form:

$$
\begin{pmatrix}
1 & 0.2 & 0.06 & 0.024 & 0.0108 & 0.0054 \\
0.2 & 1 & 0.3 & 0.12 & 0.054 & 0.027 \\
0.06 & 0.3 & 1 & 0.4 & 0.18 & 0.09 \\
0.024 & 0.12 & 0.4 & 1 & 0.45 & 0.225 \\
0.0108 & 0.054 & 0.18 & 0.45 & 1 & 0.5 \\
0.0054 & 0.027 & 0.09 & 0.225 & 0.5 & 1
\end{pmatrix}.
$$

For this structure only specification of the unstructured matrix would be correct. The AR(1) structure would require equal off-diagonal elements of the correlation matrix, while the Markov structure would specify three different (and necessarily unequal) structures, because the timing of measurements differs between structures.

The true correlation structures for our simulations were therefore AR(1), Markov, or decaying product, while the working correlation structures were AR(1), Markov, and unstructured. As mentioned above, the working unstructured matrix is correct for the AR(1) and decaying product structure structures, but not for the Markov structure due to the lack of common timings for participants. Work is currently underway (Shults) to implement QLS for the decaying product structure.

8.2   Simulation Results

Here we discuss the results for the simulations that were conducted according to the scenarios described in the previous section.

8.2.1   True AR(1) Structure

The simulation results for the true AR(1) structure are provided in Table 8.1. The unstructured matrix was selected most often by every criteria except for QIC_A; however, this was not an incorrect choice. The $RJ1$, $RJ2$, and $DBAR$ approaches might be considered to have had the best performance, because they almost always selected a correct structure. The QIC and QIC_A criteria incorrectly selected the identity structure 16.42 and 43.04 percent of the time, respectively.

If we consider the total number of criteria that selected each structure in each simulation run, the identity structure was selected by zero criteria 57.0 percent of the time, the exchangeable structure was chosen by zero criteria 97.0 percent of the time, and the Markov correlation structure was selected by zero criteria 88.9 percent of the time. Therefore, the incorrectly specified exchangeable and Markov structures, and to a lesser degree the incorrectly specified identity matrix, were unlikely to be selected by any criteria. (When the identity structure was incorrectly selected, this was largely due to incorrect selection by the QIC or QIC_A criteria.)

The exchangeable, identity, and Markov structures were selected by three or more criteria in less than 3 percent of simulation runs. The AR(1) structure was selected by no criteria 19.1 percent of the time; however, in each of these runs, the correct unstructured matrix was selected by three or more criteria. In contrast, the AR(1) structure was selected by three or more criteria 32.6 percent of the time, while the unstructured matrix was selected by three or more criteria 71.67 percent of the time. This suggests that a structure should be ruled out if it is not selected by any criteria. Or, if a particular working structure is selected by three or more criteria, this provides strong evidence that this structure is a correct structure.

8.2.2   True Markov Structure

The simulation results for the true Markov structure are provided in Table 8.2. The unstructured matrix is incorrect for this scenario because the timings were not the same for all subjects, so that the true structure varied between subjects, while the unstructured matrix assumes that the correlation structure is the same for all subjects. The AR(1) structure was only correct for a subset of subjects (23 of 138).

The Markov structure was selected most often by the QIC, QIC_A, and CIC criteria. Every criteria except for QIC_A selected the unstructured matrix more than 25 percent of the time, while $DBAR$ chose the unstructured matrix slightly more than

Table 8.1 *Percentage of 1,000 Simulation Runs in Which the AR(1), Unstructured (UNS), Identity, Exchangeable (EXC), or Markov Matrices were Selected for the RJ1, RJ2, DBAR, QIC, QIC$_A$, and CIC Criteria. The data were simulated using the model described in Section 8.1 for a true AR(1) structure; however, the unstructured structure is also correct for this scenario. The percentage pertaining to the structure that was selected most often is in boldface. The percentages for correctly specified structures are underlined.*

Criteria	AR(1)	UNS	Identity	EXC	Markov
RJ1	42.64	**56.56**	0	0	0.8
RJ2	42.14	**57.86**	0	0	0
DBAR	25.53	**74.47**	0	0	0
QIC	15.92	**62.36**	16.42	1	4.3
QIC$_A$	**48.15**	0	43.04	2.2	6.61
CIC	10.11	**81.98**	2.6	1	4.3

half the time. The *QIC* and *QIC$_A$* incorrectly selected the identity structure in 7.4 and 34.6 percent of simulation runs, respectively.

If we consider the total number of criteria that selected each structure in each simulation run, the identity structure was selected by no criteria 65.4 percent of the time, the exchangeable structure was chosen by no criteria 95.4 percent of the time, and the AR(1) structure was chosen by no criteria 58.3 percent of the time. Therefore, the incorrectly specified identity, exchangeable, and AR(1) (and especially the exchangeable) structures were unlikely to be incorrectly selected by any criteria. (However, the incorrect unstructured matrix was selected by zero criteria only 19.0 percent of the time, and by one and two criteria 24.6 percent and 22.1 percent of the time, respectively.)

The exchangeable and identity structures were selected by three or more criteria in only 1 percent of simulation runs. The likelihood of selection by three or more criteria was also very low (3.5 percent) for the AR(1) structure, but was 34.3 percent for the incorrect unstructured matrix. In contrast, the correct Markov structure was selected by three or more criteria in 62.3 percent of simulation runs. This again suggests that a structure should be ruled out if it is not selected by any criteria. Or, if a particular working structure is selected by three or more criteria, this provides strong evidence that this structure is a correct structure. However, incorrect selection by three or more criteria is more likely for the unstructured matrix (at 34.3 percent) than for the other incorrectly specified structures (at less than 3 percent).

8.2.3   True Decaying Product Structure

The simulation results for the true decaying product structure are provided in Table 8.3. The decaying product structure was selected by every criteria most often, with the exception of the *QIC$_A$* criterion, while the AR(1) structure was selected next most often. The *DBAR* structure had perhaps the best performance, while the *CIC* also performed well. Again the *QIC* and *QIC$_A$* criteria selected the identity structure a relatively high proportion of times (11.1 and 47.1 percent of the time, respectively).

Table 8.2 *Percentage of 1,000 Simulation Runs in Which the AR(1), Unstructured (UNS), Identity, Exchangeable (EXC), or Markov Matrices were Selected for the RJ1, RJ2, DBAR, QIC, QIC$_A$, and CIC Criteria. The data were simulated using the model described in Section 8.1 for a Markov structure. Only the Markov structure is correct for this scenario, although the AR(1) structure is correct for a subset of subjects. The percentage pertaining to the structure that was selected most often is in boldface.*

Criteria	AR(1)	UNS	Identity	EXC	Markov
RJ1	28.9	**38.9**	0	0.4	31.8
RJ2	17.6	**42.2**	0	0	40.2
DBAR	3.2	**53.5**	0	.1	43.2
QIC	6.1	26.8	7.4	3.4	**56.3**
QIC$_A$	3.8	0	34.6	2.5	**59.1**
CIC	1.1	30.1	1	1.9	**65.9**

Table 8.3 *Percentage of 1,000 Simulation Runs in Which the AR(1), Unstructured (UNS), Identity, Exchangeable (EXC), or Markov Matrices were Selected for the RJ1, RJ2, DBAR, QIC, QIC$_A$, and CIC Criteria. The data were simulated using the model described in Section 8.1 for a decaying product correlation structure. Only the unstructured matrix was correct for this scenario. The percentage pertaining to the structure that was selected most often is in boldface.*

Criteria	AR(1)	UNS	Identity	EXC	Markov
RJ1	37.5	**62.5**	0	0	0
RJ2	32.1	**67.9**	0	0	0
DBAR	11.2	**88.6**	0	0	0
QIC	22.6	**62.1**	11.1	1.0	3.2
QIC$_A$	**45.6**	0	47.1	2.9	4.4
CIC	9.7	**85.7**	1.5	0.8	2.3

If we consider the total number of criteria that selected each structure in each simulation run, the identity structure was selected by no criteria 52.9 percent of the time (with any incorrect selections largely due to the *QIC* and *QIC$_A$* criteria), the exchangeable structure was chosen by no criteria 96.2 percent of the time, the Markov structure was chosen by no criteria 93.1 percent of the time, and the AR(1) structure was chosen by zero criteria 21.5 percent of the time. Therefore, the incorrectly specified identity, exchangeable, and Markov structures were unlikely to be incorrectly selected by any criteria.

The exchangeable, identity, and Markov structures were selected by three or more criteria in less than 3 percent of all simulation runs. The likelihood of selection by three or more criteria was also low (23.8 percent) for the AR(1) structure. In contrast, the correct unstructured structure was selected by three or more criteria in 83.4 percent of simulation runs. This again suggests that a structure should be ruled out if it is not selected by any criteria. Or, if a particular working structure is selected by three or more criteria, this provides strong evidence that this structure is a correct structure.

8.3   Summary and Recommendations

We considered several criteria for selection of a working correlation structure for QLS and GEE. Our assessment involved product correlation structures that were not considered in earlier comparisons of criteria for selection of a working structure. Our simulation results suggest that for selection of a working correlation structure for QLS (or GEE) analysis of longitudinal binary data, when the true structure was AR(1) or decaying product unstructured, the *RJ* and *CIC* criteria had the best performance. For the true Markov structure, the *QIC*, QIC_A, and *CIC* were most likely to correctly select the true structure. However, the *QIC* and QIC_A were the only criteria that had a tendency to incorrectly select the identity structure. (The identity structure is a special case of all structures, but is not the correct choice unless the correlations are zero.)

Like some previous authors, we also observed a tendency to select the unstructured matrix over a simpler form. The QIC_A criterion that was penalized for a larger number of correlation parameters reduced the tendency of the *QIC* to select the unstructured matrix over a simpler form; however, it also increased the likelihood of incorrect selection of an identity structure. In addition, the QIC_A criterion did not select the unstructured correlation matrix, even when it was the only correct structure. However, as the sample size is increased, the QIC_A tends to the *QIC* criterion in value; therefore, the performance of both measures will be more similar for larger sample sizes. We also note that if we replace $p + 1$ with p in both the numerator and denominator of (8.5), then the performance of the QIC_A criterion is almost unchanged.

If we selected the structure for which the *RJ*1 and *RJ*2 were smallest, instead of choosing the structure with smallest distance from 1, then this markedly improved their ability to correctly specify the Markov structure (from 31.8 percent to 62.9 percent for *RJ*1, and from 40.2 to 63.0 percent for the *RJ*2 criterion). However, further study is needed to explain whether there is a theoretical justification for our observation that choosing smaller values sometimes resulted in superior performance, in comparison with choosing values that are closer to one. It would also be of interest to provide stronger theoretical justification for the adjusted *QIC* criterion, QIC_A, analogous to the proof that the bias corrected *AIC* criterion, AIC_C, is asymptotically efficient for autoregressive models (Hurvich and Tsai, 1989, Appendix).

The *RJ* criteria are especially appealing because they are based on the intuitively reasonable observation that the model based and sandwich based covariance matrices should be close in value when the working correlation structure is correctly specified. From a practical standpoint, it is helpful when the results are similar for the model and sandwich-based covariance matrix. For example, sometimes the results are only significant for the sandwich matrix; in this situation it is difficult to suggest application of the sandwich covariance matrix as the more conservative choice. It would also be helpful to assess the fitted values of the RJ criteria, for the final selected structure. For example, if the estimated values of $(RJ1, RJ2, DBAR)$ are far from $(1,1,0)$ in value for all working correlation structures, then this might suggest that the true structure was not included in the list of candidate structures.

However, our simulations suggested that it would be reasonable to calculate the *RJ*1, *RJ*2, *DBAR*, *QIC*, *QIC$_A$* and *CIC* criteria, perhaps giving more weight to the *RJ* criteria (*RJ*1, *RJ*2, *DBAR*) and *CIC* criteria than to the *QIC* or *QIC$_A$* criteria. One should be especially cautious of selecting an identity structure on the basis of the *QIC* or *QIC$_A$* criteria alone, because they were the only criteria that had an incorrect tendency to choose the identity structure. As mentioned earlier, previous authors also showed a tendency of the QIC to incorrectly select the identity structure. The assessment of existing methods, and development of new methods, for selection of a working correlation structure for QLS and GEE is still an open area of research.

In addition, it is especially important to consider the totality of results. For example, if a structure is not selected by any criteria, then this is strongly suggestive that this particular structure should be ruled out for further consideration as the final working structure. In addition, if a structure is selected by a majority of criteria, then this should be viewed as strong evidence that this is a correct structure. For analysis of binary data, as discussed in the previous chapter, we should also rule out any correlation structure that results in a severe violation of the Prentice constraints (Shults et al., 2009).

As mentioned earlier, our simulation results were almost identical for QLS and GEE, so that our suggestions are applicable for both QLS and GEE. Another approach that more naturally enables model selection and that provides a goodness-of-fit test for the validity of the first moment assumption is the method of quadratic-inference functions (*QIF*) (Qu and Song, 2004). Development of new approaches and evaluation of existing criteria for goodness-of-fit and selection of an appropriate working structure will continue to be important open areas of research for both QLS and GEE.

8.4   Exercises

Exercise 8.1 *GEE is often praised for being robust to the choice of working correlation structure. Explain in what sense GEE is robust to specification of the wrong working structure; is the same also true for QLS?*

Exercise 8.2 *Implementation of the sandwich-based covariance matrix is often suggested as a means of guarding the results of an analysis against misspecification of the true correlation structure. Are there any drawbacks to this approach? In particular, will the sandwich covariance matrix of $\hat{\beta}$ always be a consistent estimator of the covariance matrix of $\hat{\beta}$ when the true correlation structure has been incorrectly specified in the analysis? (Hint: See Sutradhar and Das (1999, 2000) for more discussion of this topic.)*

Exercise 8.3 *Some prior researchers complained that particular criteria for selection of a working correlation structure did not do well in distinguishing between the identity, exchangeable, and AR(1) structures, when the true structure was identity. Was this criticism warranted?*

Exercise 8.4 *Suppose we are planning a simulation study to compare several criteria for selection of a working correlation structure in a QLS analysis. Suppose the true structure is a 3×3 AR(1) structure with a positive correlation parameter α. For what value of α would we expect the methods to perform worst with respect to distinguishing between the true AR(1) and a working exchangeable structure?*

Exercise 8.5 *Some prior researchers complained that particular criteria for selection of a working correlation structure did not do well in distinguishing between the exchangeable and unstructured structures, when the true structure was exchangeable. Was this criticism warranted?*

Exercise 8.6 *In Section 5.5.2 we presented results of an analysis for which the choice of correlation structure made a difference. Use the criteria presented in this chapter to compare the fit of the Markov versus AR(1) structures in that example.*

Chapter 9

Sample Size and Demonstration

Sample size calculation and estimation of power are crucial first steps in planning a new study. When we plan longitudinal analyses for a particular study design, we might determine the required number of subjects to detect a meaningful difference between groups, for assumed values for the type 1 error, power, and number of measurements per subject. For cross-sectional clustered data, we might determine the total number of clusters for an assumed number of measurements per cluster. We sometimes refer to the sample size as the required number of subjects, or clusters, although we should keep in mind that the total sample size is the number of subjects multiplied by the assumed number of measurements per subject, or the number of clusters multiplied by the assumed number of measurements per cluster. For example, in a study that collects information on subjects at baseline and at 6 and 12 months post baseline, if we determine that we need 30 subjects, this means that we will collect a total of $30 \times 3 = 90$ measurements.

Methods for power and sample size are well developed for independent measurements; for example, release 13.0 of Stata includes an extensive suite of commands for assessment of sample size and power (Stata Press, 2013). When planning a study that will yield longitudinal or clustered data, the standard approaches for independent measurements must be amended to account for the correlation within a subject, or cluster. This chapter describes approaches for correlated data that are equally applicable for QLS and GEE and are based on findings by Shih (1997), who provided formulae that utilize the asymptotic covariance matrix of $\sqrt{m}(\widehat{\boldsymbol{\beta}} - \boldsymbol{\beta})$, that was provided in Equation (2.23). Sample size formulae were also provided in Diggle et al. (2002). If we assume that the true correlation structure $T_i(\rho)$ for subject i is correctly specified and is equal to $R_i(\alpha)$, then Equation (2.23) simplifies as follows:

$$V_{\boldsymbol{\beta}} = \phi \lim_{m \to +\infty} m \left(\sum_{i=1}^{m} D_i' V_i^{-1} D_i \right)^{-1}, \qquad (9.1)$$

where $D_i = \frac{\partial \mu_i}{\partial \boldsymbol{\beta}} = \left(\frac{\partial \mu_{ij}}{\partial \beta_k} \right)$ and $V_i = A_i^{1/2} R_i(\alpha) A_i^{1/2}$, as defined in Section 2.3.1.

In this chapter we will consider the situation in which we specify canonical link functions, for which $D_i = \frac{\partial \mu_i}{\partial \boldsymbol{\beta}} = A_i X_i$; substitution into Equation (9.1) then yields

175

the following simplified version of the asymptotic covariance matrix:

$$V_\beta = \phi \lim_{m \to +\infty} m \left(\sum_{i=1}^{m} E_i \right)^{-1}, \tag{9.2}$$

where

$$E_i = X_i' A_i^{1/2} R_i^{-1}(\alpha) A_i^{1/2} X_i. \tag{9.3}$$

The sample size and power formulae are based on the asymptotic covariance matrix (9.2), which depends on ϕ, $A_i^{1/2}$, $R_i(\alpha)$, and X_i.

We consider the following null versus alternative hypotheses:

$$H_O : \beta_p = 0 \text{ versus} \tag{9.4}$$
$$H_A : \beta_p = \Delta > 0, \tag{9.5}$$

where β_p is the last parameter of the p-dimensional regression parameter β.

Then, a two-sided test of H_O with type I error rate γ and type II error rate B (or equivalently, power = $1 - B$) requires the following number of subjects (or clusters) (Shih, 1997):

$$m = \frac{V_\beta[p,p](Z_{\gamma/2} - Z_{1-B})^2}{\Delta^2} \tag{9.6}$$

$$= \frac{S[p,p](Z_{\gamma/2} - Z_{1-B})^2}{(\Delta/\sqrt{\phi})^2}, \tag{9.7}$$

where $V_\beta[p,p]$ is the (p,p)-th element of V_β, $Z_\gamma = \Phi^{-1}(1-\gamma)$ is the $100(1-\gamma)$-th percentile of the standard normal distribution, and $S[p,p]$ is the is the (p,p)-th element of

$$S = 1/\phi V_\beta \tag{9.8}$$

$$= \lim_{m \to +\infty} m \left(\sum_{i=1}^{m} E_i \right)^{-1} \tag{9.9}$$

where E_i is defined in Equation (9.3).

The expression in Equation (9.7) is a function of the effect size:

$$\text{effect-size} = \Delta/\sqrt{\phi}, \tag{9.10}$$

which has a well-understood meaning for certain study designs and continuous outcomes ($\phi \neq 1$). The assumed number of measurements per subject, n, will also impact the required number of subjects, through its influence on the value of $V_\beta[p,p]$.

Alternatively, given the number of subjects (or clusters) m, the power is

$$1 - B = 1 - \Phi\left(Z_{\gamma/2} - \frac{\Delta\sqrt{m}}{\sqrt{V_\beta[p,p]}}\right) \tag{9.11}$$

$$= 1 - \Phi\left(Z_{\gamma/2} - (\Delta/\sqrt{\phi})\sqrt{\frac{N}{S[p,p]}}\right). \tag{9.12}$$

9.1 Two-Group Comparisons

In medical studies, we are often interested in comparing two groups. For example, the goal might be to compare the average pain levels between two treatment groups when repeated pain scores are available on each subject. Or, interest might focus on a comparison of the *change* in pain levels over time between two groups.

Suppose that P is the proportion of subjects in group one, so that there are Pm subjects in group one, and $(1-P)m$ subjects in group two. Here we consider scenarios in which E_i in Equation (9.3) takes two possible values, so that $E_i = E_1$ for the Pm subjects in group one, and $E_i = E_2$ for the $(1-P)m$ subjects in group two. Then, Equation (9.2) can be simplified as follows:

$$V_\beta = \phi \lim_{m \to +\infty} m(PmE_1 + [1-P]mE_2)^{-1} \tag{9.13}$$

$$= \phi \lim_{m \to +\infty} m\frac{1}{m}(PE_1 + [1-P]E_2)^{-1} \tag{9.14}$$

$$= \phi(PE_1 + [1-P]E_2)^{-1}. \tag{9.15}$$

We will consider two situations: when we are interested in making a time-averaged comparison of two group means and when we want to compare change over time in an outcome between two groups. We will consider continuous and binary outcomes.

9.1.1 Two-Group Comparisons

9.1.1.1 Time-Averaged Comparison of Group Means

We first consider a continuous outcome and the following model that allows for a comparison of group means, in a cross sectional study, or in a longitudinal study that compares the "time-averaged" mean outcome between groups.

$$E(y_{ij}) = g^{-1}(\delta_{ij}), \tag{9.16}$$

where

$$\delta_{ij} = \delta_i \tag{9.17}$$

$$= \beta_0 + \beta_1 I(\text{group one}); \tag{9.18}$$

$g^{-1}(\delta) = \delta$; $I(\text{group one}) = 1$ for subjects in group one (and 0 for subjects in group two); $j = 1,\ldots,n$; and $i = 1,\ldots,m$. We also assume that $\beta_1 > 0$ so that group one is assumed to have the larger group mean. The number of subjects in group one is Pm and the number of subjects in group two is $(1-P)m$, where P is the proportion of subjects in group one. In addition, when we consider the Markov structure, we assume that the timings on all subjects are $t_1, t_2, \ldots, t_n$.

Then, E_i in Equation (9.3) can be expressed for subjects in group one, for whom

$$X_1 = (1_n, 1_n), \tag{9.19}$$

as

$$E_1 = 1_n' R^{-1}(\alpha) 1_n \begin{pmatrix} 1 & 1 \\ 1 & 1 \end{pmatrix}, \tag{9.20}$$

where 1_n is an $n \times 1$ vector of ones. The first column in Equation (9.19) corresponds to the constant β_0 in model (9.18), while the second column corresponds to the value of one for the variable $I(\text{group one})$ for subjects in group one. Furthermore, E_i in Equation (9.3) can be expressed for subjects in group two, for whom

$$X_2 = (1_n, 0_n), \tag{9.21}$$

as

$$E_2 = 1_n' R^{-1}(\alpha) 1_n \begin{pmatrix} 1 & 0 \\ 0 & 0 \end{pmatrix}, \tag{9.22}$$

where 0_n is an $n \times 1$ vector of zeros that corresponds to the value of the indicator variable, $I(\text{group one})$, for subjects in group two.

Next, substitution into Equation (9.15) and straightforward calculations (Shih, 1997) yields the following expression for V_β: (See Exercise 9.1.)

$$V_\beta = \frac{\phi}{(1_n' R^{-1}(\alpha) 1_n) \, P(1-P)} \begin{pmatrix} P & -P \\ -P & 1 \end{pmatrix}, \tag{9.23}$$

so that

$$V_\beta[2,2] = \frac{\phi}{(1_n' R^{-1}(\alpha) 1_n) \, P(1-P)}. \tag{9.24}$$

Then, substitution into Equations (9.6) and (9.7) indicates that a two-sided test of $H_O : \beta_1 = 0$ with type I error rate γ and type II error rate B (or equivalently, power = $1 - B$) requires the following number of subjects (or clusters) (Shih, 1997):

$$m = \frac{(Z_{\gamma/2} - Z_{1-B})^2}{\phi \, (1_n' R^{-1}(\alpha) 1_n) \, P(1-P) \Delta^2} \tag{9.25}$$

$$= \frac{(Z_{\gamma/2} - Z_{1-B})^2}{(\Delta/\sqrt{\phi})^2 \, (1_n' R^{-1}(\alpha) 1_n) \, P(1-P)}. \tag{9.26}$$

Table 9.1 *Values of* $1_n'R^{-1}(\alpha)1_n$ *When the Correlation Structure* $R(\alpha)$ *is Exchangeable,* *AR(1), and Markov. For the Markov structure,* $C_{j,j+1} = \alpha^{t_{j+1}-t_j}$.

Structure	$1_n'R^{-1}(\alpha)1_n$
Exchangeable	$\dfrac{n}{1+(n-1)\alpha}$
AR(1)	$\dfrac{(n-2)\alpha^2-2(n-1)\alpha+n}{1-\alpha^2}$
Markov	$\dfrac{1}{1-C_{1,2}^2}+\sum_{k=2}^{n-1}\dfrac{1-C_{k-1,k}^2 C_{k,k+1}^2}{(1-C_{k-1,k}^2)(1-C_{k,k+1}^2)}-2\sum_{k=1}^{n-1}\dfrac{C_{k,k+1}}{1-C_{k,k+1}^2}+\dfrac{1}{1-C_{n-1,n}^2}.$

Alternatively, given the total number of subjects (or clusters) size m, the power is

$$1-B \;=\; 1-\Phi\left(Z_{\gamma/2}-\delta\sqrt{\frac{m\left(1_n'R^{-1}(\alpha)1_n\right)P(1-P)}{\phi}}\right) \tag{9.27}$$

$$=\; 1-\Phi\left(Z_{\gamma/2}-(\Delta/\sqrt{\phi})\sqrt{m\left(1_n'R^{-1}(\alpha)1_n\right)P(1-P)}\right). \tag{9.28}$$

The sample size and power formulae given in Equations (9.25) and (9.27) depend on the true correlation structure through the value of the quadratic form, $1_n'R^{-1}(\alpha)1_n$, which is the sum of all values of the elements of $R^{-1}(\alpha)$. Table 9.1 displays the values of $1_n'R^{-1}(\alpha)1_n$ for the exchangeable, AR(1), and Markov structures.

The sample size that is calculated using Equation (9.25) must often be rounded up to the nearest whole number. For example, if the required group size is 80.2 at 80 percent power, this number must be rounded up to 81 subjects per group, for which the power should then slightly exceed 80 percent. Conversely, because the number of subjects is an integer, it might not be possible to achieve 80 percent power exactly. For example, the achieved power might be 79.2 percent for $m = 30$, but 81.1 percent for $m = 31$.

Next, Table 9.2 displays the required group sizes ($m/2$) for a two-group comparison of group means, assuming the power is 80 percent, or 90 percent, to detect effect sizes of 0.25 and 0.50. For a comparison of two means, the effect size represents the absolute value of the difference in group means, divided by the standard deviation of the outcome variable. For planning purposes, it is ideal to base anticipated effect sizes on some preliminary data, or published results from another study. However, as mentioned earlier, a benefit of the effect size for a simple comparison of group means is that it has an interpretation that is well understood by many grant reviewers.

Table 9.2 shows that we do not require as many subjects when we increase the number of measurements. However, although we double the sample size when we double the number of measurements per subject, we do not necessarily cut the required number of subjects in half. For example, the first two lines of Table 9.2 indicate that if we double the number of measurements per subject from $n = 3$ to $n = 6$, the required number of subjects per group drops from 134 to 105, in order to detect an effect size of 0.25 with 80 percent power when $\alpha = 0.3$ and the true correlation

Table 9.2 *Group Size (m/2) Required for Model (9.18), for a Two-Group Time-Averaged Comparison of Group Means for a Continuous Outcome and Equal Group Sizes. The assumed power is 80 percent and 90 percent. Group sizes are displayed for effect sizes (ES) of 0.25 and 0.50; n = number of measurements per subject = 3 and 6; true correlation parameter α = 0.3 and 0.6; and true correlation structures = exchangeable (EXC), AR(1), and Markov (MARK). The timings for the Markov structure are $(1, 2, 5, 9, 11, 18)$ for n = 6 and are $(1, 2, 5)$ for n = 3.*

ES	n	α	Power = 80%			Power = 90%		
			EXC	AR(1)	MARK	EXC	AR(1)	MARK
0.25	3	0.3	134	121	102	180	162	136
0.25	6	0.3	105	69	48	141	92	64
0.5	3	0.3	34	31	26	45	41	34
0.5	6	0.3	27	18	12	36	23	16
0.25	3	0.6	185	168	133	247	225	178
0.25	6	0.6	168	112	62	225	150	83
0.5	3	0.6	47	42	34	62	57	45
0.5	6	0.6	42	28	16	57	38	21

structure is exchangeable. The number of subjects is more than halved when we double the number of measurements per subject for the Markov structure, but this is because the timings are $(1, 2, 5, 9, 11, 18)$ for this structure; the assumed study duration is therefore 18 time units, which is three times as long as the study duration for the AR(1) structure. In addition, the follow-up time for the Markov structure is longer in the second half of the study than in the first half, so that we are more than doubling the follow-up time when we increase the sample size from $n = 3$ to $n = 6$ for this structure.

The required group sizes are largest for the exchangeable structure, so that the exchangeable structure is the most conservative choice *for a comparison of time-averaged group means*. For a time-averaged comparison of group means, the required sample size also increases within increasing α. This is intuitively reasonable. Suppose the measurements are identical within clusters, so that we have perfect correlation within clusters; in this situation, the effective sample size would simply be the number of clusters, because the additional number of measurements per cluster would not yield additional information about the value of the group mean.

9.1.1.2 Time-Averaged Comparison of Proportions

We next consider model (9.16) for a time-averaged comparison of two group means, but for binary outcomes, with $g^{-1}(\delta) = exp(\delta)/(1 + exp(\delta))$. Recall that for binary variables, $E(y_{ij}) = \text{pr}(y_{ij} = 1) = p_{ij}$ and $q_{ij} = 1 - p_{ij}$. The test of hypothesis $H_O : \beta_1 = 0$ is equivalent to the test of $H_O : p_1 = p_2$, where p_1 is the probability of a positive response in group one, and p_2 is the probability of a positive response in group two. Again, we assume that $\beta_1 > 0$, so that group one is assumed to have the higher probability of success.

The derivations for comparison of two time-averaged proportions are almost the same as for the comparison of two time-averaged group means in the previous section, with two differences. First, $\phi = 1$ in Equation (9.23). Next, $A_i = \text{diag}(p_{i1}q_{i1}, \ldots, p_{in}q_{in})$ in Equation (9.3), where $p_{ij} = p_1$ for subjects in group one, and $p_{ij} = p_2$ for subjects in group two. Straightforward calculations indicate that a two-sided test of $H_O : p_1 = p_2$ with type I error rate γ and type II error rate B (or equivalently, power $= 1 - B$) requires the following total number of subjects (or clusters) (Shih, 1997): (See Exercise 9.2.)

$$m = \frac{(Z_{\gamma/2} - Z_{1-B})^2}{P_{AVE} \Delta^2 \mathbf{1}_n' R^{-1}(\alpha) \mathbf{1}_n}, \qquad (9.29)$$

where

$$P_{AVE} = \frac{1}{p_1(1-p_1)P} + \frac{1}{p_2(1-p_2)(1-P)}, \qquad (9.30)$$

$\Delta = \ln(p_1(1-p_2)/((1-p_1)p_2))$ is the log odds-ratio of response in group one relative to group two, P is the proportion of subjects in group one, and $1 - P$ is the proportion in group two. Alternatively, given the sample size m, the power is

$$1 - B = 1 - \Phi\left(Z_{\gamma/2} - \Delta\sqrt{mP_{AVE} \mathbf{1}_n' R^{-1}(\alpha) \mathbf{1}_n}\right). \qquad (9.31)$$

Required sample sizes are not as easy to summarize for proportions as they are for means. This is because the required sample size depends on the particular values of the proportions p_1 and p_2. For example, for a comparison of means, we might say that we have 80 percent power to detect an effect-size of 0.50; this effect size could correspond to an infinite variety of group means, as long as their absolute difference, divided by $\sqrt{\phi}$, equals 0.50. In contrast, the sample size calculation for proportions does not just depend on the log-odds ratio of the two proportions, but varies according to the actual values of the specified proportions.

9.1.1.3   Comparison of Change over Time for Continuous Outcomes

We next consider Model (9.16), but with δ_{ij} defined so as to allow for a comparison of change over time in a continuous outcome between two groups (Jung and Ahn, 2003; Tu et al., 2004):

$$\delta_{ij} = \delta_i \qquad (9.32)$$
$$= \beta_0 + \beta_1 I(\text{group one}) + \beta_2 \text{visit} + \beta_3 I(\text{group one}) \times \text{visit}. \qquad (9.33)$$

We assume that $\beta_3 > 0$. In addition, when we consider the Markov structure, we assume $visit = t_1, t_2, \ldots, t_n$.

Then, E_1 in Equation (9.3) can be expressed for subjects in group one with

$$X_1 = (\mathbf{1}_n, \mathbf{1}_n, \mathbf{v}_n, \mathbf{v}_n), \qquad (9.34)$$

where $\mathbf{v}_n$ is an $n \times 1$ vector that contains the ordered values of $visit$. The first column in (9.34) corresponds to the constant β_0 in Equation (9.33); the second column corresponds to the value of one for the indicator variable I(group one) for subjects in group one; the third column contains the ordered visit values; and the fourth column contains the values of the group by visit interaction term for group one. Furthermore, E_2 in (9.3) can be expressed for subjects in group two by

$$E_1 = (\mathbf{1}_n, \mathbf{0}_n, \mathbf{v}_n, \mathbf{0}_n),\tag{9.35}$$

where $\mathbf{0}_n$ in the second and fourth column corresponds to the value of the indicator variable, I(group one), and group by visit interaction term, respectively, for subjects in group two.

The asymptotic covariance matrix V_β can then be calculated by substituting Equations (9.34) and (9.35) into Equation (9.15). Then, Equations (9.6) and (9.7) can be used to calculate the sample size and power for a a two-sided test of Equation (9.4) with $\beta_p = \beta_3$, that is, of the hypothesis $H_O : \beta_3 = 0$ with type I error rate γ and type II error rate $\mathbf{B}$ (or equivalently, power = 1 - $\mathbf{B}$).

9.1.1.4   Comparison of Change over Time for Binary Outcomes

We next consider model (9.16) with δ_{ij} as defined in Equation (9.32) and $g^{-1}(\delta) = exp(\delta)/(1 + exp(\delta))$.

The derivations for comparison of change over time in a binary outcome between two groups are the same as those presented in the previous section, but with two differences: $\phi = 1$ in Equation (9.23), and the A_i in Equation (9.3) are as defined in Section 9.1.1.2.

Again, the sample sizes for comparison of binary outcomes over time depend on the particular value of the specified proportions. It is then very easy to demonstrate that, for comparison of change over time between groups, increasing correlation leads to reduced sample sizes. In addition, application of the AR(1) structure is the more conservative choice because application of the AR(1) structure requires a larger sample size in this testing situation than does the exchangeable structure.

9.2   More Complex Situations

The two-group comparisons presented in the previous section can easily be extended for multiple groups. Suppose, for example, that the proportion of subjects in group one is P_1; the proportion of subjects in group two is P_2; and the proportion in group three is $1 - P_1 - P_2$. Furthermore, suppose that E_i in Equation (9.3) takes value $E_i = E_1$ for the $P_1 m$ subjects in group one, $E_i = E_2$ for the $P_2 m$ subjects in group two, and $E_i = E_3$ for the $(1 - P_1 - P_2) m$ subjects in group three. As in Equation (9.13), we can then easily show that (9.2) simplifies as follows:

$$V_\beta = \phi (P E_1 + P_2 E_2 + [1 - P_1 - P_2] E_3)^{-1}.\tag{9.36}$$

However, rather than attempt to obtain simplified expressions for Equation (9.36) it might be most straightforward to now simply write a program, for example, in Stata, to calculate (9.36) for a particular scenario. In general, calculating V_β in Equation

(9.2) amounts to assuming that the study design stays constant as new patients are enrolled; we can then calculate V_{β} using the same code that we would use to obtain the model-based covariance estimator of $\hat{\beta}$, but calculated now at the assumed values for β, α, and ϕ.

Power to detect clinically relevant carry over in cross-over trials was assessed in Putt (2006). Power and sample size for cluster-unit trials was discussed in Preisser et al. (2003b), who presented an integrated approach for design, analysis, and sample size determination. The methods presented in Preisser et al. (2003b) are applicable to data with two sources of correlation (see Chapter 8), for example, if repeated measurements are collected on patients within geographically defined clusters; conversely, Teerenstra et al. (2010) describe sample size calculations for two sources of correlation in a three-level cross-sectional design, using a Kronecker product correlation structure and a closed-form expression for $\mathbf{1}_n'R^{-1}(\alpha)\mathbf{1}_n$, for which Equation (9.29) is applicable.

Sample size software GEESIZE version 3.1 is available that is based on publications by Rochon (1998) and Dahmen et al. (2004). GEESIZE computes the minimum sample size for several study designs for GEE and is therefore applicable for analyses in which QLS will be applied. GEESIZE is an SAS macro that uses SAS IML, so that SAS is required to use this software.

9.3   Worked Example

In the following section we demonstrate the implementation of QLS in the analysis of data from a controlled clinical data on the best initial therapy for treating patients with bipolar type II (BP II) depression. To facilitate sharing the data, 30 percent of measurements were randomly dropped, prior to analysis. Our goal is to demonstrate QLS in a comparison of the antidepressant efficacy of Effexor (venlafaxine) versus lithium monotherapy in patients with BP II MDE. However, a complete analysis of the treatments with respect to efficacy, and other outcomes of interest such as hypomanic and mixed mood conversions, is beyond the scope of our demonstration.

In the trial, patients were randomized to monotherapy with Effexor 37.5 to 375 mg daily or lithium 300 to 2,100 mg daily for 12 weeks. The primary outcome measure was the 28-item Hamilton Depression Rating (HAM-D 28), with embedded "typical" HAM-D 17 and "atypical" HAM-D 17-R symptom cores. For testing hypotheses regarding correlated correlations at baseline, for example to compare the correlation between age and HAM-D 17 with the correlation between age and HAM-D 28 in males versus females, Bilker et al. (2004) proposed a two-factor ANOVA-like test, CORANOVA, for correlated correlations. For our demonstration, we consider the HAM-D 17 score in the first 7 weeks of the trial. Higher HAM-D 17 scores indicate greater levels of depression, so our goal will be to evaluate both treatments with respect to their ability to yield a reduction in HAM-D 17 scores over the course of the trial.

Graphical displays of the data suggested a roughly linear relationship between HAM-D 17 group means and time, within each treatment group. Figure 9.1 displays individual-level overlaid plots of HAM-D 17 scores versus week of the study, by

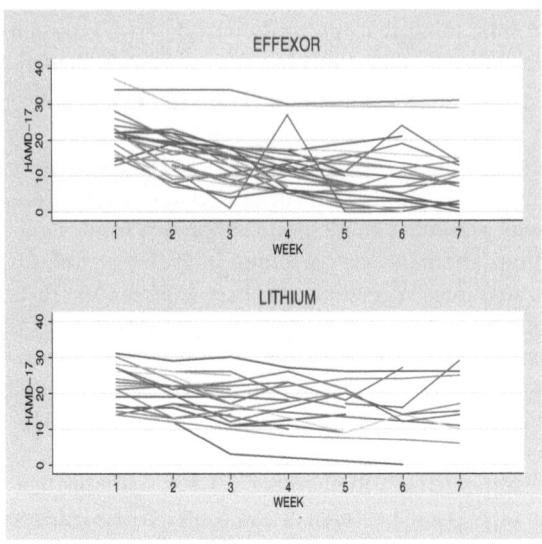

Figure 9.1 *Individual-level fitted means of HAM17 scores versus time for patients in each treatment group.*

treatment group. We therefore implemented model (9.32) with QLS to compare the change over time in HAM-D 17 scores between the two groups. Our primary hypothesis of interest was $H_o : \beta_3 = 0$, versus $H_o : \beta_3 \neq 0$, because a significant interaction would indicate that the treatment groups differ significantly with respect to change in HAM-D 17 scores over time. We tested the hypothesis with a two-sided test and a p-value < 0.05 as the criterion for statistical significance. The analysis was conducted in Stata 13.0, with the **xtqls** command (Shults et al., 2007) to implement QLS.

We first assess the number of measurements by treatment group and week of the trial.

```
.  **Sample size by visit:
.  table Effexor time, c(n ham17)
```

randomized _drug== Effexor	1	2	3	time 4	5	6	7
0	17	19	19	9	12	10	9
1	22	22	16	14	14	14	18

We see that patients dropped out of the study, and in particular in the lithium

Table 9.3 *Estimated Values of the Regression Parameters, Correlation Parameter* α, *and Scalar Parameter* ϕ, *in the Analysis of Data from a Clinical Trial to Compare Effexor versus Lithium. The results are shown for several working correlation structures: Markov, AR(1), exchangeable (EXC), and identity. The interaction parameters* β_3 *is our primary parameter of interest; therefore, the p-value for the two-sided test of* $H_o : \beta_3 = 0$ *is provided in parentheses. The predicted percentage reduction in HAM-D scores from baseline are also provided for the Effexor (PR1) and lithium (PR2) treatment groups.*

Structure	$\hat{\beta}_0$	$\hat{\beta}_1$	$\hat{\beta}_2$	$\hat{\beta}_3$	$\hat{\alpha}$	$\hat{\phi}$	PR1	PR2
Markov	21.72	0.23	−1.03	−0.94 (0.08)	0.7	49.8	59.3	29.9
AR(1)	21.85	0.07	−1.01	−1.00 (0.03)	0.66	49.7	60.5	28.9
EXC	21.54	−0.55	−1.09	−0.77 (0.02)	0.64	49.7	58.2	31.9
Identity	21.64	−0.42	−0.94	−1.02 (0.03)	0	49.5	61.1	27.2

group, with only 9 of 17 lithium patients remaining at the end of the trial. Patients also missed visits intermittently, as is indicated by the decrease and subsequent increase in sample sizes over the course of the trial. In any clinical trial, the reasons for patient drop-out must be carefully assessed. QLS and GEE assume that any missing measurements are missing completely at random (MCAR); if the reasons for drop-out suggest that this is not the case, then other methods, such as likelihood-based approaches that do not require the MCAR assumption, might be applied in a sensitivity analysis. For example, Liao et al. (2012) implemented a likelihood-based approach with a transition probability model for missingness, in order to analyze continuous longitudinal responses with non-ignorable, non-monotone missing data. Guo et al. (2004) presented a pattern mixture model for longitudinal data with informative dropout. For further discussion of power analysis for longitudinal and clustered studies with missing data, see Roy et al. (2007) and Tu et al. (2007).

We anticipated that HAM-D scores that were measured more closely together on a patient, might be more similar than those that were farther apart in time. For this reason, the AR(1) and Markov structures were considered to be most plausible, with the Markov structure being the first choice because it takes the unequal temporal spacing of measurements for some patients into account. However, we also fit the identity and exchangeable structures to assess the sensitivity of results to choice of correlation structure. (Application of the unstructured matrix resulted in a failure to converge for GEE.) We have seen the output of a QLS analysis in earlier chapters, and therefore summarized the results of this analysis in Table 9.3.

Table 9.3 indicates that the reduction in HAM-D 17 scores was significantly greater for Effexor, for every fitted structure except the Markov structure, for which the p-value for the test of $H_o : \beta_3 = 0$ (versus $\beta_3 \neq 0$) was 0.08. The estimated percentage reductions were also calculated as 100 times the difference between the final and initial scores, divided by the initial HAM-D scores. The estimated percentage reductions were 58.2 to 61.1 percent for patients treated with Effexor, versus 27.2 to 31.9 percent for patients with lithium. These reductions are clinically meaningful, especially for those treated with Effexor. Figure 9.2 displays the fitted means for the Markov structure. They indicate a similar pattern that was also observed for the other

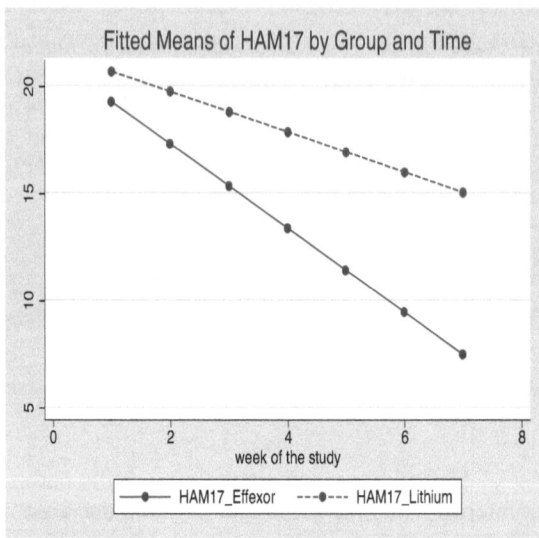

Figure 9.2 *Fitted means of HAM17 scores versus time for patients in each treatment group.*

structures; the groups were similar at baseline, with a greater reduction over time in HAMD-17 scores for patients treated with Effexor. The estimated correlations were also quite large, and ranged in value from 0.64 to 0.70 for the non-identity structures.

We next assess the fit of the different correlation structures; this is especially important because we did not achieve statistical significance for the Markov structure. Table 9.4 displays the estimated criteria for assessment of fit of the various structures. Table 9.4 indicates that the Markov structure was selected by the *RJ*1 and *RJ*2 criteria, while the AR(1) structure was selected by the *DBAR* criterion. The identity was selected by the *QIC* and *QICA*, while the exchangeable structure was selected by the *CIC* criterion.

Because no structure was selected by a majority of criteria, we did not have a clear winner with respect to selection of a best-fitting correlation structure. However, as we observed in our simulations in Chapter 8, the *QIC* and *QICA* both tended to over-select the identity structure. We would therefore be fine with ruling out the identity structure, given the large estimated value of α for the other structures and the poor performance of the identity structure with respect to the *RJ*1, *RJ*2, and *DBAR* criteria (with respective values of 2.3, 7.36, and 3.76, when 1, 1, and 0 indicate good fit).

Our selection of a working structure was important because our results depend on the choice of working structure. Given the plausibility of the Markov structure and the fact that it was identified as having the best fit by the *RJ* criteria, we were not comfortable in ruling out this structure. Our analysis therefore suggests that there is a beneficial impact of Effexor relative to lithium in the treatment of BP II depression. However, further studies are needed to confirm this result because we did not achieve

Table 9.4 *Estimated Values of Criterion for Selection of Working Structure for the Depression Study. Estimated values for the RJ1, RJ2, DBAR, QIC, QIC$_A$, and CIC criteria are displayed for each working structure. The structure that is selected by each criterion is displayed in boldface.*

Structure	RJ1	RJ2	DBAR	QIC	QICA	CIC
Markov	**0.99**	**1.23**	0.25	11498.06	11498.43	8.60
AR(1)	1.03	1.24	**0.17**	11487.85	11488.22	8.43
EXC	1.37	2.13	0.40	11458.99	11459.36	**8.38**
Identity	2.3	7.36	3.76	**11427.41**	**11427.67**	9.19

statistical significance according to our predetermined criterion for significance for the Markov structure.

9.3.1 Sample Size for a Future Study

Here we demonstrate sample size calculations for a future study that are informed by the results for the depression study. We use the user-written Stata command **compareslope** (Shults) to obtain the sample size for a true AR(1), exchangeable, and identity structure. For purpose of the demonstration, we assume the parameter values are the estimated values for the working AR(1) structure in Table 9.3. (Some of the output is removed below.) The command **compareslope** takes the following arguments: α, β_0, β_1, β_2, β_3, ϕ, n, and P (the proportion in the intervention group). For purposes of demonstration, we assume that $n = 3$, so that the timings are 0, 1, and 2 weeks on each patient.

The sample size calculations were developed for a test of $H_o : \beta_3 = 0$ versus $H_o : \beta_3 = \Delta > 0$, so we provide a positive value for β_3. (The values for β_0, β_1, and β_2 will not have an impact on the results.)

```
. compareslope  .66  21.92    -.07  -2.01       1  49.65  3 .5
Sample Size Calculations for a comparison of slopes. We will
input beta0, beta1, beta2, but they do not impact the sample
size calculations. Only the coefficient for the interaction
term matters. This assumes the timings are 0, 1, 2, ...
which has an impact on the results.

Assumed alpha, beta0, beta1, beta2, beta3, tau, n, P (% in
treatment group),effect-size

assumed[1,9]
     alpha     BO     B1     B2     B3    phi     n    P   ES
r1     .66  21.92   -.07  -2.01      1  49.65     3   .5  .14

Effect size, n, alpha, Required Sample Size for Intervention and
Control Group for AR(1),  Exchangeable, Identity:
```

```
results[3,7]
          ES        n    alpha   NInt80   NCon80   NInt90   NCon90
AR1      .14        3     .66      220      220      295      295
EXC      .14        3     .66      133      133      178      178
 Id      .14        3      0       390      390      522      522
```

The above output shows that the estimated effect size in the preliminary study was 0.14. For a true exchangeable structure, effect size of 0.14, n = 3, $\alpha = 0.66$, and timings = (0, 1, 2) weeks on each patient, the required group sizes in the intervention and control groups are 133 for power = 80 percent and are 178 for power = 90 percent. It also shows that the required sample sizes are greater for an AR(1) structure, so that an assumption of the AR(1) structure is more conservative than an assumption of the exchangeable structure for a comparison of two slopes.

Next, we decrease the value of α to 0.30.

```
. compareslope   .3   21.92    -.07   -2.01   1   49.65   4  .5

Assumed alpha, beta0, beta1, beta2, beta3, tau, n,
  P (% in treatment group),effect-size


assumed[1,9]
     alpha      B0      B1      B2      B3     phi       n      P      ES
r1      .3   21.92    -.07   -2.01      1   49.65       4     .5     .14
```

Effect size, n, alpha, Required Sample Size for Intervention and Control Group for AR(1), Exchangeable, Identity:

```
results[3,7]
          ES        n    alpha   NInt80   NCon80   NInt90   NCon90
AR1      .14        4     .3       166      166      222      222
EXC      .14        4     .3       110      110      147      147
 Id      .14        4      0       156      156      209      209
```

We see that the required sample sizes increases when the value of α is decreased. For a comparison of two slopes, the required sample size will decrease for larger effect sizes, number of measurements per patient, and values of α. For a comparison of time-averaged group means, the results would be the same, with one exception; for time-averaged group means, the required sample sizes increase for larger values of α.

9.4   Discussion and Summary

Sample size and power calculations are extremely important when planning medical studies. In fact, some of our medical colleagues are most likely to reach out to a statistician when they want to figure out how many subjects "they need" in order to conduct their study. If the sample size is too large, then money will be wasted and treatment differences that are statistically significant, but not clinically meaningful, could be identified. On the other hand, a sample size that is too small could result in loss of power to detect meaningful group differences. In addition to thoughtful determination of the sample size, Ellenberg (1994) discussed the importance of carefully assessing the method of sample selection in order to prevent reporting spurious associations. Ellenberg (1994) presented several examples of selection bias in research that led to results that were wrong or misleading. Ellenberg (2000) discussed practical issues involved in the conduct of clinical trials with placebo or active control groups.

We largely focused on two-group comparisons, which are extremely important in clinical trials. We considered assessment of time-averaged group means, and change over time between groups. Some important differences between the two types of comparisons are as follows. For time-averaged group means, the exchangeable structure is the most conservative choice, because its application in sample size calculations will require the largest group sizes. In addition, assuming higher values for the correlation results in larger required group sizes for comparison of time-averaged means. For comparison of change over time between two groups, as assessed via group by time interaction, the results are reversed. Application of the exchangeable structure requires smaller sample sizes relative to the AR(1) structure, while larger correlations result in smaller required sample sizes. In each situation, assuming the most plausible structure (exchangeable for time-averaged group means and AR(1) for comparison of change over time) is the most conservative choice for the true correlation structure.

We discussed power and sample size calculations based on the asymptotic covariance matrix, but simulations might also be used to reassure ourselves that the results do indeed hold for a finite sample. Another advantage of simulation studies is that they allow for assessment of anticipated power in different situations. Simulations are straightforward to conduct for continuous outcome variables using the multivariate normal distribution. However, the use of simulation to assess power for discrete variables with patterned correlation structures is hindered by the lack of an analog to the multivariate normal distribution for discrete data. Guerra and Shults (2013) demonstrated the use of simulations to assess power for GEE for discrete data that are overdispersed relative to the Poisson or binomial distributions. Over-dispersion is a relatively common feature of count data, so it was of interest to evaluate the potential loss in power that occurs when the over-dispersion is ignored.

In general, the development of methods to evaluate power and calculate sample size typically lag behind the development of new statistical approaches; as a result, proposals for studies often describe a simple method for sample size calculation, and a more complex approach for the analysis. The power for the study might then be more, or less, than what is anticipated by the investigators, because the method on which the sample size calculations were based does not match the method that will be applied in the analysis. The continued development of new methods for simulation of discrete correlated data, and of new methods for sample size calculation, remain active areas of research where useful contributions can still be made. It is important to appropriately power longitudinal studies because their findings can have important implications, for example to inform policy changes as described in French and Heagerty (2008).

9.5 Exercises

Exercise 9.1 *Simplify V_β for the time-averaged comparison of group means in Model (9.16); this will yield Equation (9.23) and Equation (9.24).*

Exercise 9.2 *Simplify V_β for the time-averaged comparison of group means in Model (9.16) but for binary outcomes, with $g^{-1}(\delta) = exp(\delta)/(1 + exp(\delta))$.*

Exercise 9.3 *Suppose that we would like to plan a larger future study on the basis of the analysis results in Section 9.3.*

Determine the sample size that would be needed to achieve 80% power to detect the difference that was observed in the clinical trial in section 9.3.1, for the AR(1), exchangeable, and identity structures. Assume the timings are $0, 1, \ldots, 5$ for each patient.

Bibliography

Akaike, H. (1973). Information theory and an extension of the maximum likelihood principle. In Petrov, B. and Csaki, F., editors, *Second International Symposium on Information Theory*, pages 267–281. Akademiai Kiado, Budapest, Hungary.

Bahadur, R. (1961). A representation of the joint distribution of responses to *n* dichotomous items. In Solomon, H., editor, *Studies in Item Analysis and Prediction, Stanford Mathematical Studies in the Social Sciences*, volume VI. Stanford University Press, Stanford, California, USA.

Barnett, A., Koper, N., Dobson, A., Schmiegelow, F., and Manseau, M. (2010). Using information criteria to select the correct variance covariance structure for longitudinal data in ecology. *Methods in Ecology and Evolution*, 1(1):15–24.

Bellamy, S., Gibberd, R., Hancock, L., Howley, P., Kennedy, B., Klar, N., Lipsitz, S., and Ryan, L. (2000). Analysis of dichotomous outcome data for community intervention studies. *Statistical Methods in Medical Research*, 2:135–160.

Bellamy, S., Li, Y., Lin, X., and Ryan, L. (2005). Quantifying pql bias in estimating cluster-level covariate effects in group-randomized trials. *Statistica Sinica*, 15(4):1015–1032.

Bilker, W., Brensinger, C., and Gur, R. (2004). A two factor ANOVA-like test for correlated correlations: CORANOVA. *Multivariate Behavioral Research*, 39(4):565–594.

Breitung, J., Chaganty, N., Daniel, R., Kenward, M., Lechner, M., Martus, P., Sabo, R., Wang, Y., and Zorn, C. (2010). Discussion of "generalized estimating equations: Notes on the choice of the working correlation matrix." *Methods of Information in Medicine*, 49(5): 426–432.

By, K., Qaqish, B., Pressier, J., Perin, J., and Zink, R. (2011). ORTH: R and SAS software for regression models of correlated binary data based on orthogonalized residuals and alternating logistic regressions. *Computer Methods and Programs in Biomedicine* (in press).

Carey, V., Zeger, S., and Diggle, P. (1993). Modelling multivariate binary data with alternating logistic regressions. *Biometrika*, 80(3): 517–526.

Chaganty, C. and Mav, D. (2007). Estimation methods for analyzing longitudinal data occurring in biomedical research. In Khattree, R. and Naik, D., editors, *Computational Methods in Biomedical Research*, pages 371–400. Chapman and Hall/CRC Press, Boca Raton, Florida, USA.

Chaganty, N. (1997). An alternative approach to the analysis of longitudinal data via generalized estimating equations. *Journal of Statistical Planning and Inference*, 63(1): 39–54.

Chaganty, N. and Deng, Y. (2007). Ranges of measures of association for familial binary variables. *Communications in Statistics—Theory and Methods*, 36(3): 587–598.

Chaganty, N. and Joe, H. (2004). Efficiency of generalized estimating equations for binary responses. *Journal of the Royal Statistical Society: Series B (Statistical Methodology)*, 66(4): 851–860.

Chaganty, N. and Joe, H. (2006). Range of correlation matrices for dependent Bernoulli random variables. *Biometrika*, 93(1): 197–206.

Chaganty, N. and Naik, D. (2002). Analysis of multivariate longitudinal data using quasi-least squares. *Journal of Statistical Planning and Inference*, 103(1): 421–436.

Chaganty, N. and Shults, J. (1999). On eliminating the asymptotic bias in the quasi-least squares estimate of the correlation parameter. *Journal of Statistical Planning and Inference*, 76(1-2): 145–161.

Christensen, R. (1987). *Plane Answers to Complex Questions: The Theory of Linear Models*. Springer-Verlag Inc., New York, New York, USA.

Christensen, R. (1997). *Linear Models for Multivariate, Time Series, and Spatial Data*. Springer-Verlag Inc., New York, New York, USA.

Crowder, M. (1995). On the use of a working correlation matrix in using generalised linear models for repeated measures. *Biometrika*, 82(2): 407–410.

Cui, J. (2007). QIC program and model selection in GEE analyses. *Stata Journal*, 7(2): 209–220.

Dahmen, G., Rochon, J., Konig, I. R., and Ziegler, A. (2004). Sample size calculations for controlled clinical trials using generalized estimating equations (GEE). *Methods of Information in Medicine*, 43(5): 451–461.

Davis, C. (2002). *Statistical Methods for the Analysis of Repeated Measurements*. Springer-Verlag Inc., New York, New York, USA.

De Backer, M., De Keyser, P., De Vroey, C., and Lesaffre, E. (1996). A 12-week treatment for dermatophyte toe onychomycosis: terbinafine 250 mg/day vs. itraconazole 200 mg/day—a double blind comparative trial. *British Journal of Dermatology*, 134:16–17.

Diggle, P., Heagerty, P., Liang, K., and Zeger, S. (2002). *Analysis of Longitudinal Data, Second Edition*. Springer-Verlag Inc., New York, New York, USA.

Diggle, P., Liang, K., and Zeger, S. (1994). *Analysis of Longitudinal Data, First Edition*. Springer-Verlag Inc., New York, New York, USA.

Dobson, A. and Barnett, A. (2008). *An Introduction to Generalized Linear Models, Third Edition*. Chapman and Hall/CRC Press, Boca Raton, Florida, USA.

Dunlop, D. (1994). Regression for longitudinal data: A bridge from least squares regression. *The American Statistician*, 48(4): 299–303.

Ellenberg, J. H. (1994). Selection bias in observational and experimental studies. *Statistics in Medicine*, 13(5-7): 557–567.

Ellenberg, S. S. (2000). Placebo-controlled trials and active control trials in the evaluation of new treatments. *Annals of Internal Medicine*, 133: 464–470.

Feng, R., McClure, L. A., Tiwari, H. K., and Howard, G. (2009). A new estimate of family disease history providing improved prediction of disease risks. *Statistics in Medicine*, 28(8):1269–1283.

Fitzmaurice, G. (1995). A caveat concerning independence estimating equations with multivariate binary data. *Biometrics*, 51(1): 309–317.

French, B. and Heagerty, P. J. (2008). Analysis of longitudinal data to evaluate a policy change. *Statistics in Medicine*, 27(24): 5005–5025.

Galecki, A. (1994). General class of covariance structures for two or more repeated factors in longitudinal data analysis. *Communications in Statistics*, 23: 3105–20.

Gantz, B., Tyler, R., Knutson, J., Woodworth, G., Abbas, P., McCabe, B., Hinrichs, J., Tye-Murray, N., Lansing, C., Kuk, F., and Brown, C. (1988). Evaluation of five different cochlear implant designs: Audiologic assessment and predictors of performance. *Laryngoscope*, 98(10):1100–1106.

Godambe, V. (2002). *Estimating Functions*. Oxford University Press, Oxford, UK.

Grady, J. and Helms, R. (1995). Model selection techniques for the covariance matrix for incomplete longitudinal data. *Statistics in Medicine*, 13(14):1397–1416.

Guerra, M. and Shults, J. (2013). On the simulation of longitudinal discrete data with specified marginal means and first-order antedependence. *The American Statistician (tentatively accepted)*.

Guerra, M. and Shults, J. (forthcoming 2014). *Logistic Regression for Longitudinal Data*. Spinger, New York, New York, USA.

Guerra, M., Shults, J., Amsterdam, J., and Ten-Have, T. (2012). The analysis of binary longitudinal data with time-dependent covariates. *Statistics in Medicine*, 10: 931–948.

Guo, W., Ratcliffe, S. J., and TenHave, T. T. (2004). A random pattern-mixture model for longitudinal data with dropouts. *Journal of the American Statistical Association*, 99: 929–937.

Halekoh, U., Højsgaard, S., and Yan, J. (2006). The R package geepack for generalized estimating equations. *Journal of Statistical Software*, 15(2):1–11.

Hammill, B. and Preisser, J. (2006). A SAS/IML software program for GEE and regression diagnostics. *Computational Statistics and Data Analysis*, 51(2):1197–1212.

Hardin, J. and Hilbe, J. (2003). *Generalized Estimating Equations*. Chapman and Hall/CRC Press, Boca Raton, Florida, USA.

Hardin, J. and Hilbe, J. (2007). *Generalized Linear Models and Extensions, Second Edition*. Stata Press, College Station, Texas, USA.

Hardin, J. and Hilbe, J. (2012). *Generalized Estimating Equations, Second Edition*. Chapman and Hall/CRC Press, Boca Raton, Florida, USA.

Hedeker, D. and Gibbons, R. (2006). *Missing Data in Longitudinal Studies*, pages 279–312. John Wiley and Sons, Inc, New York, New York, USA.

Heitjan, D. F. and Basu, S. (1996). Distinguishing "missing at random" and "missing completely at random". *The American Statistician*, 50(3): 207–213.

Hilbe, J. (2009). *Logistic Regression Models*. Chapman and Hall/CRC Press, Boca Raton, Florida, USA.

Hin, L. and Wang, Y. (2009). Working-correlation-structure identification in generalized estimating equations. *Statistics in Medicine*, 28: 642–658.

Horton, N. J., Bebchuk, J. D., Jones, C. L., Lipsitz, S. R., Catalano, P. J., Zahner, G. E. P., and Fitzmaurice, G. M. (1999). Goodness-of-fit for GEE: An example with mental health service utilization. *Statistics in Medicine*, 18(2): 213–222.

Huber, P. (1967). The behavior of maximum likelihood estimation under nonstandard conditions. In *Proceedings of the Fifth Berkeley Symposium on Mathematical Statistics and Probability, Berkely, California, USA*, pages 221–233.

Hurvich, C. and Tsai, C.-L. (1989). Regression and time series model selection in small samples. *Biometrika*, 76(2): 297–307.

Jung, S. and Ahn, C. (2003). Sample size estimation for GEE method for comparing slopes in repeated measurements data. *Statistics in Medicine*, 22(8):130–515.

Kauermann, G. and Carroll, R. (2001). A note on the efficiency of sandwich covariance matrix estimation. *Journal of American Statistical Association*, 96:1387–1396.

Kim, H., Hilbe, J., and Shults, J. (2008a). On the designation of the patterned associations for longitudinal Bernoulli data: Weight matrix versus true correlation structure?. *UPenn Biostatistics Working Papers. Working Paper 29*, 26.

Kim, H. and Shults, J. (2010). %QLS SAS macro: A SAS macro for analysis of longitudinal data using quasi-least squares. *Journal of Statistical Software*, 35(2):1–22.

Kim, H. and Shults, J. (2013). Analysis of unbalanced simultaneously clustered and longitudinal data using quasi-least squares in SAS. *Under review*.

Kim, H., Shults, J., Patterson, S., and Goldberg-Alberts, R. (2008b). Analysis of adverse events in drug safety: A multivariate approach using stratified quasi-least squares. *UPenn Biostatistics Working Papers. Working Paper 29*.

Kumanyika, S., Wadden, T., Shults, J., Fassbender, J., Brown, S., Bowman, M., Brake, V., West, W., Frazier, J., Whitt-Glover, M., Kallan, M., Desnouee, E., and Wu, X. (2009). Trial of family and friend support for weight loss in African-American adults. *Archives of Internal Medicine*, 169(19):1795–1804.

Lefkopoulou, M., Moore, D., and Ryan, L. (1989). The analysis of multiple correlated binary outcomes: Application to rodent teratology experiments. *Journal of the American Statistical Association*, 84: 810–815.

Liang, K. and Zeger, S. (1986). Longitudinal data analysis using generalized linear models. *Biometrika*, 73(1):13–22.

Liang, K., Zeger, S., and Qaqish, B. (1992). Multivariate regression analysis for categorical data. *Journal of Royal Statistical Society, Series B*, 54(1): 3–40.

Liao, K., Freres, D. R., and Troxel, A. B. (2012). A transition model for quality-of-life data with non-ignorable non-monotone missing data. *Statistics in Medicine*, 31(28): 3444–3466.

Little, R. and Rubin, D. (2002). *Statistical Analysis with Missing Data (Second Edition)*. John Wiley and Sons, Inc, New York, New York, USA.

Lu, B., Preisser, J., Qaqish, B., Suchindran, C., Bangdiwala, S., and Wolfson, M. (2007). A comparison of two bias-corrected covariance estimators for generalized estimating equations. *Biometrics*, 63(3): 935–941.

Lu, N. and Zimmerman, D. (2005). The likelihood ratio test for a separable covariance matrix. *Statistics and Probability Letters*, 73: 449–457.

Magnus, P., Gjessing, H., Skrondal, A., and Skjaerven, R. (2001). Paternal contribution to birth weight. *Journal of Epidemiology and Community Health*, 55: 873–877.

Mancl, L. and DeRouen, T. (2001). A covariance estimator for GEE with improved small-sample properties. *Biometrics*, 57(1):126–134.

McCullagh, P. and Nelder, J. (1989). *Generalized Linear Models, Second Edition*. Chapman and Hall/CRC Press Monographs on Statistics and Applied Probability, Boca Raton, Florida, USA.

Miller, M., Davis, C., and Landis, J. (1993). The analysis of longitudinal polytomous data: Generalized estimating equations and connections with weighted least squares. *Biometrics*, 49(4):1033–1044.

Molenberghs, G. (2010). Editorial. "Generalized estimating equations: Notes on the choice of the working correlation matrix". *Methods of Information in Medicine*, 49(5): 419–420.

Molenberghs, G. and Kenward, M. (2010). Semi-parametric marginal models for hierarchical data and their corresponding full models. *Computational Statistics & Data Analysis*, 54(2): 585– 597.

Molenberghs, G. and Verbeke, G. (2005). *Models for Discrete Longitudinal Data*. Springer, New York, New York, USA.

Monahan, J. (2008). *A Primer on Linear Models*. Chapman and Hall/CRC Press, Boca Raton, Florida, USA.

Naik, D. and Rao, S. (2001). Analysis of multivariate repeated measures data with a Kronecker product structured covariance matrix. *Journal of Applied Statistics*, 28: 91–105.

Nelder, J. and Wedderburn, R. (1972). Generalized linear models. *Journal of the*

Royal Statistical Society, Series A, 135(3):370–384.

Newton, H. (1988). *TIMESLAB: A Time Series Analysis Laboratory*. Wadsworth and Brooks/Cole, Pacific Grove, Califormia, USA.

Nunez-Anton, V. and Woodworth, G. (1994). Analysis of longitudinal data with un-equally spaced observations and time-dependent correlated errors. *Biometrics*, 50(2): 445–456.

O'Brien, S., Holubkov, R., and Reis, E. (2004). Identification, evaluation, and man-agement of obesity in an academic primary care. *Pediatrics*, 114(e):154–159.

Olkin, I. and Pratt, J. W. (1958). A multivariate Tchebycheff inequality. *The Annals of Mathematical Statistics*, 29(1): 226–234.

Pan, W. (2001). Akaike's information criterion in generalized estimating equations. *Biometrics*, 57:120–125.

Park, T. (1993). A comparison of the generalized estimating equation approach with the maximum likelihood approach for repeated measurements. *Statistics in Medicine*, 12(18):1723–1732.

Preisser, J., Arcury, T., and Quandt, S. (2003a). Detecting patterns of occupational illness clustering with alternating logistic regressions applied to longitudinal data. *American Journal of Epidemiology*, 158(5): 495–501.

Preisser, J., Qaqish, B., and Perin, J. (2008). A note on deletion diagnostics for estimating equations. *Biometrika*, 95(2): 509–513.

Preisser, J. and Qaqish, B. F. (1996). Deletion diagnostics for generalised estimating equations. *Biometrika*, 83(3): 551–562.

Preisser, J., Reboussin, B., Song, E., and Wolfson, M. (2007). The importance and role of intracluster correlations in planning cluster trials. *Epidemiology*, 18(5): 552–560.

Preisser, J., Young, M., Zaccaro, D., and Wolfson, M. (2003b). An integrated population-averaged approach to the design, analysis, and sample size deter-mination of cluster-unit trials. *Statistics in Medicine*, 22(8): 1235–1254.

Preisser, J. S., By, K., Perin, J., and Qaqish, B. F. (2012). Deletion diagnostics for alternating logistic regressions. *Biometrical Journal*, 54(5): 701–715.

Preisser, J. S., Lohman, K., and Rathouz, P. (2002). Longitudinal studies with dropouts missing at random. *Statistics in Medicine*, 21: 3035–3054.

Prentice, R. (1988). Correlated binary regression with covariates specific to each binary observation. *Biometrics*, 44(4): 1033–1048.

Putt, M. E. (2006). Power to detect clinically relevant carry-over in a series of cross-over studies. *Statistics in Medicine*, 25(15): 2567–2586.

Qaqish, B. (2003). A family of multivariate binary distributions for simulating correlated binary variables with specified marginal means and correlations. *Biometrika*, 90: 455–463.

Qaqish, B. and Liang, K. (1992). Marginal models for correlated binary responses with multiple classes and multiple levels of nesting. *Biometrics*, 48(3): 939–

950.

Qaqish, B., Zink, R., and Preisser, J. (2012). Orthogonalized residuals for estimation of marginally specified association parameters in multivariate binary data. *Scandinavian Journal of Statistics*, 39(3): 515–527.

Qu, A. and Li, R. (2006). Quadratic inference functions for varying coefficient models with longitudinal data. *Biometrics*, 62(2): 379–391.

Qu, A. and Lindsay, B. (2003). Building adaptive estimating equations when inverse of covariance estimation is difficult. *Journal of the Royal Statistical Society, Series B*, 65(1):127–142.

Qu, A., Lindsay, B., and Li, B. (2000). Improving generalised estimating equations using quadratic inference functions. *Biometrika*, 87(4): 823–836.

Qu, A. and Song, P. (2004). Assessing robustness of generalised estimating equations and quadratic inference functions. *Biometrika*, 91(2): 447–459.

Qu, Y., Williams, G., Beck, G., and Medendorp, S. (1992). Latent variable models for clustered dichotomous data with multiple subclusters. *Biometrics*, (4):1095–1102.

Ratcliffe, S. and Shults, J. (2008). GEEQBOX: A MATLAB toolbox for implementation of quasi-least squares and generalized estimating equations. *Journal of Statistical Software*, 25(14):1–13.

Rencher, A. and Schaalje, G. (2007). *Linear Models in Statistics, Second Edition*. Wiley-Interscience, Hoboken, New Jersey, USA.

Rochon, J. (1998). Application of GEE procedures for sample size calculations in repeated measures experiments. *Statistics in Medicine*, 17(14):1643–1658.

Rotnitzky, A. and Jewell, N. (1990). Hypothesis testing of regression parameters in semiparametric generalized linear models for cluster correlated data. *Biometrika*, 77: 485–497.

Roy, A., Bhaumik, D., Subhash, S., and Gibbons, R. (2007). Sample size determination for hierarchical longitudinal designs with differential attrition rates. *Biometrics*, 63(3): 699–707.

Roy, A. and Khattree, R. (2003). Tests for mean and covariance structures relevant in repeated measures-based discriminant analysis. *Journal of Applied Statistical Science*, 12(2): 91–104.

Roy, A. and Khattree, R. (2005). On implementation of a test for Kronecker product covariance structure for multivariate repeated measures data. *Statistical Methodology*, 2: 297–306.

Rubin, D. B. (1976). Inference and missing data. *Biometrika*, 63(3):581–592.

Sabo, R. and Chaganty, N. (2009). Adaptation of quasi-least squares to estimate correlations within a nuclear family. *Communications in Statistics: Theory and Methods*, 38(16): 3059–3076.

Sabo, R. and Chaganty, N. (2010a). Estimation methods for an autoregressive familial correlation structure. *Communications in Statistics: Theory and Methods*,

39(6):973–991.

Sabo, R. and Chaganty, N. (2010b). Hypothesis testing for various familial dependence structures. *Communications in Statistics: Simulation and Computation*, 13(1): 207–219.

Shaw, P. A. and Proschan, M. A. (2013). Null but not void: Considerations for hypothesis testing. *Statistics in Medicine*, 32(2):196–205.

Shi, G. and Chaganty, N. (2004). Application of quasi-least squares to analyze replicated autoregressive time series regression models. *Journal of Applied Statistics*, 31(10):1147–1156.

Shih, W. (1997). Sample size and power calculation for periodontal and other studies with clustered samples using the method of generalized estimating equations. *Biometrical Journal*, 39(8):899–908.

Shults, J. (1996). *The Analysis of Unbalanced and Unequally Spaced Longitudinal Data Using Quasi-Least Squares*. PhD thesis, Department of Mathematics and Statistics, Old Dominion University, Norfolk, Virginia, USA.

Shults, J. (2011). Generalized estimating equations: Notes on the choice of the working correlation matrix - continued. Letter to the editor. *Methods of Information in Medicine*, 50(1): 96–99.

Shults, J. and Chaganty, N. (1998). Analysis of serially correlated data using quasi-least squares. *Biometrics*, 54(4):1622–1630.

Shults, J., Mazurick, C., and Landis, J. (2006a). Analysis of repeated bouts of measurements in the framework of generalized estimating equations. *Statistics in Medicine*, 25(23): 4114–4128.

Shults, J. and Morrow, A. (2002). Use of quasi-least squares to adjust for two levels of correlation. *Biometrics*, 58(3): 521–530.

Shults, J. and Ratcliffe, S. (2009). Analysis of multi-level correlated data in the framework of generalized estimating equations via xtmultcorr procedures in STATA and QLS functions in MATLAB. *Statistics and Its Interface*, 2:187–196.

Shults, J., Ratcliffe, S., and Leonard, M. (2007). Improved generalized estimating equation analysis via XTQLS for implementation of quasi-least squares in Stata. *Stata Journal*, 7(2):147–166.

Shults, J., Sun, W., Tu, X., Kim, H., Amsterdam, J., Hilbe, J., and Ten-Have, T. (2009). A comparison of several approaches for choosing between working correlation structures in generalized estimating equation analysis of longitudinal binary data. *Statistics in Medicine*, 28(8): 2338–2355.

Shults, J., Tu, X., and Amsterdam, J. (2006b). On the violation of bounds for the correlation in generalized estimating equation analyses of binary data from longitudinal trials. *UPenn Biostatistics Working Papers. Working Paper 8*.

Shults, J., Whitt, M., and Kumanyika, S. (2004). Analysis of data with multiple sources of correlation in the framework of generalized estimating equations.

Statistics in Medicine, 23(20):3209–3226.

Song, P. (2007). *Correlated Data Analysis: Modeling, Analytics, and Applications.* Springer Advances in Statistics, New York, New York, USA.

Spiegelhalter, D. J., Best, N. G., Carlin, B. P., and Van Der Linde, A. (2002). Bayesian measures of model complexity and fit. *Journal of the Royal Statistical Society: Series B (Statistical Methodology)*, 64(4): 583–639.

Stokes, M., Davis, C., and Koch, G. (2000). *Categorical Data Analysis Using the SAS System, 2nd edition.* SAS Institute Inc., Cary, North Carolina.

Sugiura, N. (1978). Further analysis of the data by Akaike's information criterion and the finite corrections. *Communications in Statistics, Theory and Methods*, A(7):13–26.

Sun, W., Shults, J., and Leonard, M. (2009). A note on the use of unbiased estimating equations to estimate correlation in analysis of longitudinal trials. *Biometrical Journal*, 51(1): 5–18.

Sutradhar, B. and Das, K. (1999). On the efficiency of regression estimators in generalised linear models for longitudinal data. *Biometrika*, 86(2): 459–465.

Sutradhar, B. and Das, K. (2000). On the accuracy of efficiency of estimating equation approach. *Biometrics*, 56(2): 622–625.

Teerenstra, S., Lu, B., Preisser, J., Achterberg, T., and Borm, G. (2010). On the accuracy of efficiency of estimating equation approach. *Biometrics*, 66(10):1230–1237.

Troxel, A. B., Fairclough, D. L., Curran, D., and Hahn, E. A. (1998). Statistical analysis of quality of life with missing data in cancer clinical trials. *Statistics in Medicine*, 17(5-7): 653–666.

Troxel, A. B., Lipsitz, S. R., Fitzmaurice, G. M., Ibrahim, J. G., Sinha, D., and Molenberghs, G. (2010). A weighted combination of pseudo-likelihood estimators for longitudinal binary data subject to non-ignorable non-monotone missingness. *Statistics in Medicine*, 29(14): 1511–1521.

Tu, X., Kowalski, J., Zhang, J., K.G., L., and P., C.-C. (2004). Power analyses for longitudinal trials and other clustered designs. *Statistics in Medicine*, 23(18): 2799–2815.

Tu, X., Zhang, J., Kowalski, J., Shults, J., Feng, C., Sun, W., and Tang, W. (2007). Power analyses for longitudinal study designs with missing data. *Statistics in Medicine*, 26(15): 2958–2981.

Wang, Y. and Carey, V. (2003). Working correlation structure misspecification, estimation and covariate design: Implications for generalised estimating equations performance. *Biometrika*, 90(1): 29–41.

Wedderburn, R. (1974). Quasi-likelihood functions, generalized linear models, and the Gaussian method. *Biometrika*, 61(3): 439–447.

Weissfeld, L. and Kshirsagar, A. (1992). A modified growth curve model and its application to clinical studies. *Australian and New Zealand Journal of Statistics*,

34(2):161–168.

Westgate, P. M. (2013). A bias correction for covariance estimators to improve inference with generalized estimating equations that use an unstructured correlation matrix. *Statistics in Medicine*, 32(16): 2850–2858.

White, H. (1982). Maximum likelihood estimation of misspecified models. *Econometrica*, 50(1):1–25.

White, M. (2011). Determinants of change in vitamin D level in chronic kidney disease. Master's thesis, University of Pennsylvania, Philadelphia, Pennsylvania.

Whittle, P. (1958). A multivariate generalization of Tchevichev's inequality. *The Quarterly Journal of Mathematics*, 9(1): 232–240.

Xie, J. and Shults, J. (2009). Implementation of quasi-least squares with the R package qlspack. *UPenn Biostatistics Working Papers. Working Paper 32*.

Xie, J., Shults, J., Peet, J., Stambolian, D., and Cotch, M. (2010). Quasi-least squares with mixed linear correlation structures. *Statistics and Its Interface*, 3(2): 223–234.

Yan, J. (2002). GEEPACK: Yet another package for generalized estimating equations. *R News*, 2:12–14.

Zeger, S., Liang, K., and Albert, P. (1988). Models for longitudinal data: A generalized estimating equation approach. *Biometrics*, pages 1049–60.

Zeger, S., Liang, K., and Self, S. (1985). The analysis of binary longitudinal data with time-independent covariates. *Biometrika*, 72(3): 31–38.

Zhao, L. P. and Prentice, R. (1990). Longitudinal data analysis for discrete and continuous outcomes. *Biometrika*, 77(3): 642–648.

Zheng, B. (2000). Summarizing the goodness of fit of generalized linear models for longitudinal data. *Statistics in Medicine*, 19:1265–1275.

Ziegler, A. (2011). *Generalized Estimating Equations*. Springer, New York, Heidelberg, Dordrecht, London.

Ziegler, A. and Vens, M. (2010). Generalized estimating equations: Notes on the choice of the working correlation matrix. *Methods of Information in Medicine*, 49(5): 421–425.

Index